Being Human
in a Technological Age

Books by or about contributors to this volume also
published by the Ohio University Press:

TROY ORGAN

The Hindu Quest for the Perfection of Man

Western Approaches to Eastern Philosophy

PAUL RICOEUR

Political and Social Essays

Studies in the Philosophy of Paul Ricoeur

BEING HUMAN IN A TECHNOLOGICAL AGE

Collected and Edited by

Donald M. Borchert and David Stewart

Ohio University Press : Athens, Ohio

Library of Congress Cataloging in Publication Data
Main entry under title:
Being human in a technological age.

    Includes bibliographical references and index.
      1. Humanism—Addresses, essays, lectures.
2. Science and civilization—Addresses, essays, lectures.
3. Technology and civilization—Addresses, essays,
lectures. I. Borchert, Donald M., 1934-
II. Stewart, David, 1938-
B105.H8B44   128   79-4364
ISBN 0-8214-0399-0
ISBN 0-8214-0427-X pbk.

Contents

Contributors

LANGDON GILKEY is Shailer Mathews Professor of Theology, The University of Chicago Divinity School. Among his books are *Maker of Heaven and Earth; Shantung Compound; Religion and the Scientific Future;* and *Reaping the Whirlwind: A Christian Interpretation of History.*

SEWARD HILTNER is Professor of Theology and Personality, Princeton Theological Seminary, and consultant for the Menninger Foundation. His books include *Religion and Health; Sex Ethics and the Kinsey Reports; The Context of Pastoral Counseling* (with Karl Menninger); and *Constructive Aspects of Anxiety.*

GERALD F. KREYCHE is Professor of Philosophy and Chairman, Department of Philosophy, DePaul University. He is past president of the American Catholic Philosophical Association, author of numerous articles on existential philosophy, and co-editor of the Harbrace Philosophy Series.

TROY ORGAN is Distinguished Professor Emeritus of Philosophy, Ohio University. His writings on Hinduism include *Hinduism: Its Historical Development; The Self in Indian Philosophy; The Hindu Quest for the Perfection of Man;* and *Western Approaches to Eastern Philosophy.*

PAUL RICOEUR is Professor of Philosophy, the University of Paris (Nanterre) and John Nuveen Professor, The University of Chicago Divinity School. Among his many books are *Freedom and Nature; the Symbolism of Evil; Freud and Philosophy;* and *Political and Social Essays.*

R. R. WILSON is Director of the Fermi National Accelerator Laboratory near Batavia, Illinois, and is a member of the National Academy of Sciences, the American Academy of Arts and Sciences, and the American Philosophical Society. He was the first chairman of the Federation of American Scientists and received the National Medal of Science in 1973.

DANIEL YANKELOVICH is a social psychologist and a principal in the firm of Yankelovich, Skelley and White, opinion analysts. He is the author (with William Barrett) of *Ego and Instinct*, which offers a humanistic model for psychology.

Introduction

There is a lot of loose talk these days about technology, loose talk from both humanists and technologists. The difficulty stems from the fact that persons on both sides of this debate tend to deal in caricatures with little real communication taking place. Technologists tend to look at humanists as having little relevance to the demands of the real world, and they see humanists as impractical intellectuals lost in the fuzzy-headed speculations of scientific illiteracy. Humanists return the compliment by viewing technologists as one-dimensional persons having no concern for the abiding heritage of Western Civilization.

One has but to read the literature to see these two caricatures being expressed. Many technologists present themselves as unabashed apologists for technological achievement, championing technology as the new savior which will free us from all human ills. The retort from humanists is for technologists to come and be instructed by humanists in order to learn why they are hollow, unfulfilled individuals. If we must forgive the unrestrained enthusiasm of technologists celebrating their mastery over our environment, we should also overlook the inconsistency of the humanist who pounds out a diatribe against technology on an electric typewriter, jumps into a late-model automobile equipped with solid-state ignition, and rushes off to spread invectives over the local educational television station. Depending on whom you read, technology is either described as an unqualified good, as the latest stage in mankind's mastery of nature, as bearer of leisure and plenty—albeit with a modicum of harmful side-effects which, in due course, will be eliminated by further advances in technology; or conversely, technology is viewed with alarm, as the Faustian expression of our age whose credo is "if we can do it, we should do it." Technology is variously credited with overcoming disease, polluting our atmosphere and waterways, freeing humankind from want, reducing men to cogs in vast industrial machines, pushing back new frontiers of knowledge thereby eliminating human ignorance, and producing destructive power capable of wiping life from the earth.

C. P. Snow labeled this debate the battle between the two

1

cultures, and though the term has entered our vocabulary thereby diagnosing this conceptual malady, little real dialogue between the two spheres has taken place. Some have questioned whether there are just two cultures, seeing instead a whole plethora of cultures, each in its own watertight cocoon unconcerned about communicating with the rest. We will leave this debate to others and instead aim at a more modest step in the direction of opening up communication. This book is predicated on the thesis that a necessary step toward overcoming the bifurcation between the two (or more) cultures is to provide a forum for the interchange of views of what it means to be a human being in a society dominated by an advanced technology. That ours is a technological age is scarcely to be doubted. And when the smoke of battle between the warring camps ceases, if it ever does, we will probably discover that technology is neither the unmitigated blessing its proponents would suggest nor the unqualified bane its opponents aver but like most things human is an ambiguous mixture both of good and evil. This volume accordingly brings together scientists—both natural and social—theologians, and philosophers to focus on the question of humanness. It offers no definitive solutions or normative conclusions but is only a step toward such conclusions. As Confucius reminds us, however, a long journey begins with a single step.

ROBERT R. WILSON—The View from the Physical Sciences

If C. P. Snow's chasm between the two cultures is to be overcome, it probably will demand the reflections of a humanist with scientific insight or a scientist with humanistic sensitivities. The first essay in this collection is precisely from such a person. Wilson is director of Fermilab in Batavia, Illinois, and among his numerous scientific credentials is his membership in the National Academy of Sciences and the fact that he was awarded the National Medal of Science in 1973. But Wilson has also received formal training as a sculptor and has been commissioned to make sculptures at several universities. It is from the viewpoint of one who has successfully bridged the two cultures in his own person that Wilson offers us the claim that physics is a truly human enterprise.

If Wilson is correct—that physics "is a beautiful creation which has meaning for man's view of himself and his place in the world"—then why was physics ever viewed as a threat to humanness? Wilson correctly points out that the origins of physics, as well as its threatening aspects, is to be found in antiquity in the atomic theories of Democritus and Lucretius. If all reality can be explained in terms of little uncuttable (hence atomic) bits of stuff moving through the void, what then can be said to be distinctive about human nature? It is interesting to note that at the birth of physical science the distinctiveness of man became a problem, for where is human dignity, free will, and human uniqueness if we are but conglomerations of atoms moving through the void? Even Lucretius backed away from the full dehumanizing implications of his theory and suggested that as atoms fall through empty space they swerve ever so slightly thus allowing for the possibility of free will. This notable fudge factor introduced in primitive physics to save the uniqueness of mankind was virtually eliminated by the modern physicists of the seventeenth and eighteenth centuries. If the molecules blindly run, they at least run according to mathematically describable laws in a secure Newtonian universe of inexorable natural law. From the vantage point of a twentieth-century scientist, Wilson decries the "oversimplified eighteenth-century arguments" drawn from physics which reduced the world to the "bleak picture" of a "deterministic materialistic world." But we can certainly understand how it happened. Physics was doing something: explaining the motion of planets, offering insights into the nature of physical reality, and developing new machines to create a better world, with the optimistic doctrine of progress that reached its zenith in the nineteenth century and was underscored by the Darwinian view of evolution.

The error was not entirely on the side of the simplistic conclusions of humanists; it was abetted by the arrogance of some physicists. For example, that notable French astrophysicist and mathematician Pierre Simon de Laplace bravely boasted in his *Philosophical Essay on Probabilities* that if a superhuman intelligence could grasp the position of every particle in the universe at a given time and know all the forces acting on them, for such an intelligence "nothing would be uncertain and the future, as the past, would be present to its eyes." Whereas Newton believed that the solar system required periodic divine adjustment to assure its stability, Laplace provided the calculations necessary to show that variations in planetary motion were cyclical and explainable in terms of the laws of planetary motion. When Napoleon asked Laplace why there was no mention of God in his theory, Laplace replied, "I have no need of that hypothesis." Small wonder that humanists and theologians felt threatened by the science of physics.

This naive confidence in the limitless power of scientific explanation was not to go unchallenged, and the challenge came from physics itself. With the advent of quantum mechanics with its indeterminancy principle and the startling insight into the space-time continuum, physics was never quite the same. A new humility, as called for by Wilson, is more appropriate to the present state of physics. There are limits to what we can know; we must be content not with absolute certainty but with statistical probability. We cannot know with certainty the path of an electron but only the probability of its course. All the old constants—space, time, causality, predictability—were called into question, and we now know that the very observation of subatomic phenomena introduces a new factor into those very phenomena themselves.

But all this aside, Wilson makes a persuasive argument for the humanness of physics, for physics is a study pursued by human beings. Beyond this first level of compatibility between physics and humanness, Wilson sees two others—the aesthetic dimension and the creative aspect inherent in physics. The aesthetic dimension of physics is a tantalizing aspect of the discipline that is frequently ignored but is put in focus by Wilson's analysis. Mathematicians frequently speak of alternative mathematical models not in terms of their explanatory power alone, but in terms of their elegance, their beauty, and simplicity. Even the hardware of physical science has an aesthetic dimension. One has but to see the Central Laboratory Building of Fermilab rising cathedral-like from the plains of Illinois to see the aptness of Wilson's comparison of accelerator builders to the cathedral builders of the thirteenth century.

There is a Kantian echo in Wilson's further claim that the scientist and the artist are both engaged in a creative enterprise. Kant spoke of two Copernican revolutions. The first displaced the earth, and man, from the center of the universe, but the second relocated man squarely at the center of the knowing process. All knowledge is but human knowledge and is shaped by the human capacity to organize, categorize, and schematize the raw data of sensibility. Wilson suggests that physical theories are less ultimate accounts of reality than an imposition of order on that reality analogous to the order impressed on canvas and paint by the artist. As creative expressions of human beings, physics is human, and "the business of physicists," Wilson says, "should also be concerned with the poetry and drama of physics, with the beauty of its form and function." In creating the world of physics, developing techniques, and building devices, the scientist is led by intuition, one of whose beacons is beauty, rather than the pedestrian ploddings of an investigator "blindly following the so-called scientific method." If humankind would recognize these polarities—analytical reason and aesthetic intuition—as essential features of all creative activity, then Snow's two cultures would be seen as an unnecessary antagonism between the two poles of a single human enterprise.

GERALD KREYCHE—The View from Existential Philosophy

Perhaps no philosophical movement in history has been as admittedly critical of technology as is existentialism. Yet, as the next paper in the collection shows, existential philosophers have not been blindly unaware of the positive effects of technology but are rather concerned with its dehumanizing capabilities. Gerald Kreyche, a noted writer on existential philosophy, begins his analysis with the insights of the French Catholic thinker Gabriel Marcel, for whom the notion of person is central. Following Marcel, Kreyche sees much in contemporary technological society that promotes the personhood of individuals. Technology has raised standards of living, increased longevity, and elevated the expectations of workers. It has opened up possibilities for cultural and aesthetic enjoyment for those who, in earlier eras, would have been condemned to an uninteresting subsistence level of existence. Kreyche's note is a welcome corrective to the tendency of some humanists to condemn technology outright, yet Kreyche also perceives a threatening side to technology. He warns against the reduction of persons to the status of quantifiable variables and cautions against the desire on the part of the humanities and social sciences to ape the method of the natural sciences. He advises against man's willingness to exchange his humanness for the promise of increased efficiency. The problem, as Kreyche pinpoints it, is not how to eliminate technology—which is both impossible and undesirable—but how to humanize it. His vision is not of a return to a pre-technological golden age but of an advance to an era of enriched humanness made possible by new technology.

Essential to the humanization of technology is the concept of person. A person is a center of meaning and value which transcends all attempts at categorization. Accordingly, Kreyche transforms the reifying question "What is man?" into the humanizing question "Who am I?"

That his question must again be raised is itself an indication of how eroded our conception of the dignity of man has become. Appealing to Marcel's distinction between *being* and *having*, Kreyche shows how dominant the sphere of having has become in our technological society. Having itself has materialistic overtones. I *have* a car, a dog, a house; but I do not have a spouse or even a body; I am married; I am my body. Having relates to the sphere of objectivity ("How much money do you have?") and to the area of the problem ("Do you have the solution yet?"). Kreyche points out that curiosity, science, and acquaintance are of the sphere of having, whereas wonder, wisdom, and friend are of the sphere of being.

The distinction between being and having is perhaps clarified by an analogous differentiation Marcel makes between problem and mystery. A problem is that which is separate from me and confronts me as something to be overcome by an increase in knowledge. Problem deals with objectivity, and the techniques of science are the appropriate response. Mystery, in contrast, is that in which I am intensely involved. No increase in knowledge can dispel mystery. Whereas a problem is separate from me, a mystery is relational, as in the mystery of the lover's bond to the beloved. If human existence is a mystery, as Kreyche following Marcel maintains, then it cannot be understood by the techniques appropriate to the resolution of problems. Problem relates to things; mystery to persons.

Here, then, is the great bifurcation: being, mystery, and person on the one hand; having, problem, and thing on the other. The threat of Skinner's approach to understanding human behavior is not that he offers quantified data but that he collapses mystery into problem; persons into things. It should also not come as a surprise to learn that Skinner rejects any notion of human freedom, or "autonomous man" as he refers to it. The sphere of the problem is on the level of things caught in the nexus of physical causality. One who remains at this level will never confront freedom any more than a person wearing red spectacles will see anything non-red. Freedom seems incapable of "proof" in the scientific sense because proof itself is of the sphere of the problem.

Finally, the sphere of mystery includes the possibility of self-transcendence. Things are never more than they are; they have an immutable essence which provides the parameters of their reality. Persons in contrast are self-transcending, always capable of becoming more than they are. Humanness, as Heidegger suggested, is not constituted by our actualities but by our possibilities. This is only another way of reinforcing the view of person as mystery and is what Kreyche means when he observes that the concept of person "represents the unknown and will always remain in part the unknowable." This capability of self-transcendence, Kreyche insists, is that which is distinctive of our humanness. Our humanness consists in the "ability to become more than what we are. Neither animal, angel nor God can claim this."

SEWARD HILTNER—The View from Theology and Psychology

In attempting to understand humanness, Seward Hiltner brings together the insights of theology and psychology with the underlying conviction that truth in one discipline reinforces truth in the other. Hiltner's openness to science, particularly the science of psychology, is consonant with the essence of the biblical tradition: the living God toward whom faith looks and in whom faith rests. This God is the God who acts and who has a purpose which he is seeking to realize in human history with human cooperation. Because God reveals himself through what he does, it is fitting that the name for God in ancient Hebrew religion is Jahweh, which probably means "I will be what I will be." The *scopus fidei* is a dynamic God who is very much akin to person and process. Biblical faith is also dynamic; it must be open to the emergence of the new, to the unexpected developments in God's purposeful action. Such a faith demands creative insight to discern where God is operative in contemporary process and courage to participate with God in his action, with all the *Angst* and risk associated with the possibility of error and failure. It is this kind of faith that seems to undergird Hiltner's enterprise and to facilitate his openness to science.

In one of his works published over two decades ago, Hiltner articulated a view of humanness which seems to have remained the foundation for his future investigations. He observes:

> My view . . . is that every man has some capability to move in the direction of self-understanding, and that he is truly human only as

he is doing so at the level of his own ability be it great or small. Every man is capable of some self-reflectiveness. To cultivate that capacity within the limits of his courage and his ability is essential if he is to achieve stature as a human being.[1]

To move in the direction of self-understanding is to be growing, and such growth while uneven follows a pattern: a spurt upward, a plateau, and a dip backward. Within this pattern persons sort out certain features of their behavior which they come to see were developed as important and useful responses of the self to past challenges but which have become outmoded forms of behavior now that the past challenges are no longer present. The self spurts upward freeing itself from the chains of outdated responses and lives on a new plateau until a dip backward augurs further growth. Given the tenacious strength of the outdated but still protective responses, Hiltner judges that the transformation of persons is possible "only on the hypothesis that under certain conditions in life new strength is available which is not of our conscious making."[2] The source of this new strength, he suggests, is the Spirit of God acting in human life.

Human development, according to Hiltner, is capable of evaluation. First, the movement away from the bondage of outdated responses and toward the liberation of one's creative powers is a standard for an individual's adjustment. Second, beyond this standard Hiltner affirms a certain God-given order or structure of the universe against which individual growth toward freedom can be evaluated.

> Underlying all individual and cultural differences there are certain basic ideals and aspirations which men have had in many ages, and we believe these bear some relation to our human destiny. Men can be molded by their cultures, up to a point. But something which is within us will then rebel. There is something about the very structure of the universe to which we must look for a more adequate standard of personality adjustment. This is not unrelated to our deepest selves, but it is not adequately conceived when viewed only as the self.[3]

In this early work, Hiltner did not move much beyond this hint that reality harbors a normative structure of humanness. Perhaps it would not be mistaken to suggest that in his paper included in this collection Hiltner offers some of his insights concerning that normative structure. Deploying the method of theological paradox, he shows that the biblical view of humanness consists in the tension-unity of a series of polarities, and that a breaking of the paradoxical tension at any one of the polarities leads to a truncated view of humanness. What he claims, however, relative to the structure of the biblical model, he seems to believe holds true for any model of humanness which is seeking to present the truth.

In brief, Hiltner argues that an adequate view of humanness must depict man as a being in tension between his animality and his self-transcendence, between the possibility of his operational model and the apparent impossibility of its actualization, between the experience of achieving and the need to grow, and between the urge to humanize now and the recognition that the humanization lies in the future. Such, then, is the human being—a being existing in a dynamic unity of polarities who grows and develops through oscillation between the polarities in an unending quest for homeostasis.

LANGDON GILKEY—THE VIEW FROM THEOLOGY AND TECHNOLOGY

Like Seward Hiltner, Langdon Gilkey seeks meaningful interchange between theology and science, and also like Hiltner he finds the category of paradox a useful device for describing the human condition. But whereas Hiltner uses paradox to illuminate a comprehensive theological model of humanness, Gilkey's essay deploys paradox to expose the religious questions raised by the promise and peril of technology.

Gilkey's assessment of technology seems to have been shaped by his experiences during his internment in a Japanese prison camp in North China during the Second World War. The camp, located in a run-down mission compound in Shantung Province, housed about two thousand Western residents of China. The internees were not arbitrarily tortured by their Japanese captors, but they were forced to create a viable society out of extremely limited resources. That creative enterprise provided Gilkey with an *in situ* basis from which to assess technology's role in human development.

At the outset, Gilkey was so impressed with the internees' accomplishments that he concluded that the human capacity to develop the technical aspects of civilization is limitless. The fundamental human problems appeared to be material in nature—housing, toilets, food production, etc.—and resolvable by techniques and organizational skills. Sliding into modern secularity, Gilkey began to think that religion was irrelevant to the real issues of human development. He recalls, "I felt convinced that man's ingenuity in dealing with difficult problems was unlimited, making irrelevant those so-called 'deeper issues' of spiritual life with which religion and philosophy pretended to deal."[4]

Problems soon began to emerge in the new community which technological know-how could not resolve. Given the scarcity of both living space and food supply in the camp, the internees at critical moments exhibited behavior motivated by a tenacious self-interest that yielded neither to the logic of justice nor to the appeals of moral obligation. When these crises threatened the viability of the young society, Gilkey became convinced that they occurred not because of inadequate techniques but because of a breakdown in character. The trouble with his new humanism, he recalls, "was not its confidence in human science and technology" but "its naive and unrealistic faith in the rationality and goodness of the men who wielded these instruments."[5] From the striking technological success in building their new society, Gilkey had generated a flawed image of man: he had projected man as the technological inquirer and inventor who uses reason dispassionately in the conquest of natural difficulties. A more realistic appraisal of humankind required acknowledgement that the self under fire deploys reason and ideals unjustly in the service of the embattled self. He concluded that "without moral health, a community is as helpless and lost as it is without material supplies and services."[6]

In searching for the pathway to that moral health, Gilkey draws on the insights of Reinhold Niebuhr and Paul Tillich to suggest that morality arises from a person's deep center of devotion—a center which provides ultimate security and meaning—a center which yields coherence and direction to one's existence. When this center is not threatened, the self's rational and moral powers are liberated for wisdom and benevolence. But when this center—this ultimate concern—is threatened, the self experiences intense anxiety which bends reason and ideals in the service of the self, while a truly rational and moral sense of what is fair and just wanes. Gilkey concluded that only a transcendent source, a source beyond the embattled self, can function as the object of an ultimate concern which can generate the needed security and free the self to be dispassionately rational and moral. He identified that transcendent source as the God of the biblical tradition in whose eternal love security is found and in whose eternal purpose meaning is discovered.

Religion—construed in terms of ultimate concern—returned to center stage for Gilkey, but this religion was rent by deep ambiguity. All humans are religious in the sense of having an ultimate concern, but their religiousness varies from the destructive and demonic to the humanizing and beatific. The critical issue is the nature of the object to which one's ultimate concern is affixed. Upon the nature of that object hangs the demonic or the saving quality of religion.

From his internment Gilkey gained not only an abiding respect for the essential role of technology in human development but also an enduring conviction that technology *by itself* cannot generate an enduring society. Religious devotion is also essential because of its capacity to free the self for wisdom and benevolence, apart from which society disintegrates through conflict, selfishness, and moral anarchy.

In the essay he has prepared for this collection, Gilkey builds on this conviction that religion, like technology, is a sine qua non for the development and maintenance of a truly human society. He attempts to show that the ambiguities of technology raise problems whose resolution is to be found only in a transcendent ground of renewal and meaning. At the outset Gilkey sketches the philosophy of history which has been generated by the scientific and technological communities from the time of Francis Bacon onward: greater empirical knowledge yields increasing power to control, direct, and remake man and his world; expanding knowledge and power, accordingly, hold a message of promise for the future, a proclamation of progress, a celebration of the triumph of intelligence, informed inquiry, and human purpose over fate; and to the modern super-technological *homo faber*, religion appears irrelevant.

This optimistic vision of Promethean man creating himself and his world as he wills has been clouded by a new *Angst* arising out of the ambiguity of science and technology as forces in history. Having promised to free humankind for richer domains of self-realization, science and technology now seem to be creating new problems and even exacerbating human bondage. The validity of scientific knowledge and the reliability of technological skills are not at issue. What is open to radical doubt is the saving and liberating character of technology.

Such questioning of scientific knowledge and control has become intensified recently, says Gilkey, by at least two factors. First, the dehumanizing effects of technology applied to social relations and institutions has become apparent: individual unique gifts, creativity and joy, have been sacrificed to the efficiency of a common systematic effort. Paradoxically, technological society promised to free the individual from external fates—crushing work, disease and want, but in many respects it has generated a more subtle bondage of inner emptiness. Second, the ecological crisis has raised the specter of the exhaustion of earth's resources. The seemingly infinite expansion of civilization and its needs, observes Gilkey, is on a collision course with the obstinate finitude of resources. Again paradoxically, technology seems to be generating the world of scarcity from which it had promised to save mankind. Furthermore, as a more authoritarian society may emerge in the form of corporate planning and control on a world-wide scale to deal with the problems of scarcity, paradoxically technology will have brought into being not the promised liberation for multifaceted self-realization but rather a less free, less affluent, less individualistic, and less innovative world.

These ambiguities of technology, however, harbor a deeper ambiguity: the ambiguity of human freedom. Technology is one of the most vivid manifestations of human freedom: the power to make and remake the human world. Rather than facilitating the enrichment of human freedom in historical process, technology seems to be creating the conditions which will absorb that freedom into authority. Gilkey observes that, paradoxically, "the exercise of technological freedom in order to remove the fates that determine freedom from the outside has *itself* become a fate that menaces freedom." Technology, the creature of freedom, has come to reveal what Gilkey calls the demonic in history: "the way our freedom is itself estranged and strangely bound." The root of these ambiguities is neither technology nor freedom and creativity per se. Rather the fault resides in the greed and selfishness which generate a demonic use of scientific intelligence and technological power with self-destructive consequences.

Echoing Tillich's method of correlation, Gilkey probes these ambiguities further to raise the questions to which religion offers the answers. Is there, he queries, any recourse from our misuse of our creativity, any source that can liberate us from our self-destructive demonic motivations? Is there any ground for meaning in history beyond the ambiguous future we confront because of our misuse of freedom? The solution to such problems can be found only in a transcendent ground of renewal and meaning. And that is the solution which technological society must confront with all seriousness: "Modern culture in the development of its science and technology has not made religion irrelevant. It has made religious understanding and the religious spirit more necessary than ever if we are to be human—and even if we are to survive."

In the end, Gilkey advocates a blend of science and technology with religious understanding—reiterating the fundamental judgment he reached during those years of social construction and philosophical assessment in the Shantung Compound. And the religion he champions is one that is anchored in divine transcendence but also committed to concrete action in the human social context.

DANIEL YANKELOVICH—The View from the Social Sciences

The paper by Daniel Yankelovich approaches the question of the meaning of humanness in contemporary society from the vantage point of the social sciences. He shows initially that the social sciences do not speak to this question with a unified voice, for one of the central problems of the social sciences today is the task of resolving the tension between the twin claims of environment and heredity as the determining factor in human development. The environmentalists insist that to be human is to be shaped by a culture and through social interaction. The alternative view posed by the perspective of evolutionary human nature insists that persons are not infinitely plastic but possess an essential human nature which is biologically based. This nature/nurture bifurcation runs throughout the social sciences, with the environmentalists having currently the upper hand. Regardless of which of these two truths one accepts,be it the environmentalist's or the hereditarian's, one will have only a partial truth, for the task that Yankelovich sees confronting the social sciences is to integrate these two truths into a holistic perspective on the meaning of humanness.

Starting with the sociological perspective of Emile Durkheim, Yankelovich applies the Durkheimian categories to a historical survey of the past two decades in our own society, then he turns to a view of that same period from an environmentalist perspective, and finally he considers some of the pressing needs that must be met if we are to achieve a fuller meaning of humanness given the innate characteristics of our human nature.

According to Durkheim, in order for a society to be held together, three things are required: a set of commonly shared values and goals, a sense of institutional legitimacy, and an unconfused sense of self. Yankelovich's hypothesis is that for roughly two decades—from the end of the Second World War until the mid-sixties—these Durkheimian categories were fulfilled by a kind of social contract in which the overwhelming presumption was the possibility of achieving success. Success was measured both by the acquisition of material possessions and upward social mobility. During this time the majority of citizens of the United States believed that the institutions of government and business were doing a creditable job, and in addition possessed a relatively unconfused sense of family roles.

All these assumptions began to change in the mid-sixties— first on campus and then gradually throughout the rest of society. The acquisitive mode of life, with its nose-to-the-grindstone work ethic, was questioned. Alternative values appeared which stressed the quality of life and not just the meaning of success measured in terms of material acquisition. More pluralistic life styles emerged with a changing view of sexual roles and a looser attitude toward the traditional values of society. Armed with the statistics of empirical surveys, Yankelovich shows that there is currently a deep skepticism in our country regarding society's institutions. Government, business, and educational institutions have all come under question, and a certain malaise has crept into the mainstream of our social life.

What all this amounts to, according to Yankelovich, is a questioning of the social contract that prevailed for two decades, and this itself shows that the environmentalist perspective is only a partial truth. The social contract of the past attempted to fit persons into the system, rather than endeavoring to mold the system to meet personal, human needs. It is as though there is an undeniable core of human concerns which no amount of material well-being can satisfy, and it is here that the alternative perspective is needed that is offered by those who defend a refulgent core of human nature.

Like Kreyche, Yankelovich calls not for the dismantling of institutions but the humanizing of them. What is needed, he says, is a concern for individuals that goes beyond the mere ethic of self-realization. We are indissolubly bound up together in society, and if we are to be fully human, we must reinstill that sense of mutal dependency which nourishes our humanness. As Yankelovich puts it, this is a "concern for others that flows from an aspect of human nature that should not be suppressed but is innate and significant."

Yankelovich ends on a note of optimism but not of inevitability. Such a society bound together not by the Durkheimian "contract" but by a sense of interdependence is attainable, but it will take major effort to achieve it. In a world shrunk by instantaneous communication, where the destiny of one nation is linked to the destiny of all the others, such an alternation in our basic relationship to others is necessary if any of us is to achieve a full measure of humanness.

PAUL RICOEUR—THE VIEW FROM SOCIAL PHILOSOPHY

The paper by Paul Ricoeur views the question of humanness in a social and political context. A dictum stemming from Aristotle and long recognized by the Western philosophical tradition is that mankind becomes human only in society. He who has no need of society is either beast or god, Aristotle said, but he is not man. Yet it is too facile to place man simply in society and thereby assume that his humanness is necessarily enhanced, for our own epoch has known only too well the dehumanizing forces that society can unleash upon persons. Often in the name of ideology, human liberties have been suppressed, invididual choices restricted, and the person submerged in the anonymous masses manipulated by the State. Utopian schemes have similarly justified the most intense restriction of human liberty for the sake of some future, yet unrealized, classless society; terror has been invoked to bring about the end of terror, and utopia often becomes just another justification for the manipulation of individuals by violence or by the threat of violence.

Ricoeur begins with what Karl Mannheim saw as deviant attitudes toward social reality—ideology and utopia. At first glance, this pair seems to be totally opposite: ideology is only a way of justifying the existing social order, whereas utopia is a way of subverting the social order in favor of some ideal construct. Ricoeur's entire analysis, however, probes the apparent contrariety between ideology and utopia to discover their dialectical interplay. Specifically, Ricoeur discovers a negative and positive side of both ideology and utopia and insists that there are congruences of both their negative sides as well as of their positive sides.

The justification of Ricoeur's claim that ideology and utopia are two modes of social imagination which have an underlying reciprocity is made more difficult in light of the critique of ideology by Marx, who saw it as essentially the way the ruling class consolidates its hold on the masses. According to Marx, the "ruling ideas" of every epoch are presented as the only rationally valid ones but in reality are the expression of the dominant material relationships controlled by the ruling class. In short, Marx held that there is an economic basis for the ideological superstructure of society, and it was the function of communism to free the proletariat from the tyranny of the ideology of the ruling class.

Ricoeur questions the adequacy of Marx's claim that there is a causal relationship between material conditions of production and ideology and suggests that Marx opened but did not explore a more fruitful path for understanding ideology by relating it to interest—and hence to the issues of domination and legitimation. Ricoeur finds that Max Weber's treatment of domination in general and legitimation in particular fills in the gap left by Marx's analysis. In Weber's view the relation between human action and "ruling ideas" is comparable to the relation between motivation and human action. In other words, Ricoeur follows Weber's claim that the function of ideology is to reinforce the rationality of the system of authority in a way that meets its claim to legitimacy. Seen from this viewpoint, the function of ideology is a conservative one—conservative in that it provides the veneer of legitimacy and rationality to the existing social order thereby giving stability and unity to society.

The pathology to which ideology is subject is that of distortion. Borrowing a term from Marx, Ricoeur calls this the "surplus value" of ideology, or to use a term in recent vogue, we might say that ideology is susceptible to a kind of ideological "overkill" which makes it immune to any criticism of the existing social order. This ideological overkill is a form of dishonesty and dissimulation which Ricoeur sees as "grafted" onto the integrative function of ideology. Without ideology, no society would be able to survive, but in its distorted form, ideological "surplus value" distorts the legitimate function of ideology by belying the possibility of alternative forms of social integration. It is at this point that utopia has a corrective role.

Ricoeur compares the relation of utopia to society to the relation of invention to science: utopia is the imaginative construction of another kind of society which opens up new possibilities for humanness. One of the significant functions of utopia is to unmask the ideological pretensions of a distorted ideology by showing alternate ways of structuring society. By doing this, utopia is a corrective to the "surplus value" of ideological overkill which is the distortion of the integrative function of ideology. In our own time, utopian visions are corrective against what Marcel calls the spirit of abstraction, as well as a reaction against the reification of the bureaucratic state and the temptation to reduce the person to anonymity.

If the pathology of ideology is dissimulation and dishonesty, the pathology of utopia is a kind of escapism that in its futurism conceals a desire for some forgotten innocence, a past golden age or paradise lost. At this point the congruence of the twin pathologies of ideology and utopia is evident: both can become adventures in untruth, and the exaggerations of utopian extravagance are no less harmful to society than the dishonesty of ideological overkill. Indeed, in their pathological forms it is often impossible to determine whether a given mode of thought is ideological or utopian. One of the ironies of twentieth-century political life is that Marx's attack on ideology was transformed by Lenin into a new dogmatism justifying the Soviet social order which has produced its own ruling classes.

Conversely there is a congruence of the positive sides of ideology and utopia. Without an ideological superstructure, there could be no unity to society; but without utopian vision, society becomes stagnant, calcified, and develops what Ricoeur metaphorically refers to as a kind of sclerosis of the body politic.

The upshot of Ricoeur's analysis is that the tension between utopia and ideology, as twin modes of the social imagination, cannot be eliminated. In a society in which humanness is enhanced, there must be a dialectical interplay between the ideology of the existing order and alternative utopian visions. At times this tension may be strained almost to the breaking point, but the tension must be preserved if our humanness is to be augmented. Utopia may appear to lead "nowhere," but at some future time society may achieve the vision which at the moment appears impossible. Ricoeur observes that " 'nowhere' may or may not refer to the 'here and now.' But who knows whether such and such an erratic mode of existence may not prophesy the man to come?" Our own society has in the past several decades seen the acceptance of proposals which only a generation ago were denounced as utopian in the extreme. But if the utopian vision is to be efficacious, it cannot ignore the givens of the existing social order, and so the dialectical interplay of these twin modes of social imagination provides the dynamics of a society and prevents stagnation and eventual ruin.

It would be a mistake to read Ricoeur as arguing for the reinstatement of a doctrine of inevitable progress. This is far from his theme, for he recognizes that there is no easy road to the attainment of the full meaning of humanness. Anyone who has lived through the wrenching traumas of social change recognizes the truth of Ricoeur's observation that the extreme forms of social imagination themselves are often necessary to effect change. When an ideology becomes extreme, or a utopian vision schizophrenic, this itself may be the initial stage in the correction of the excess. This points up further the necessary dialectical tension that must remain between ideology and utopia as dual forms of cultural imagination. "It is," Ricoeur observes, "as though we have to call upon the 'healthy' function of ideology to cure the madness of utopia and as though the critique of ideologies can only be carried out by a conscience capable of regarding itself from the point of view of 'nowhere.' "

TROY ORGAN—THE VIEW FROM THE EAST

No discussion of the meaning of humanness would be complete if it excluded the perspective of Eastern Philosophy. Troy Organ, a student of Hinduism for many years, brings to the collection an interpretation of Hinduism's view of humanness. His perspective places him in sympathy with the Neo-Vedāntic tradition, which vigorously rejects the depreciation of the space/time world which the ancient doctrine of *māyā* (illusion) preaches. In addition, Neo-Vedāntism affirms the worth and dignity of each individual in contrast to the older view that individuality is *māyā*. Neo-Vedāntism emerges as an intensely humanistic, this-worldly demythologized version of Hinduism.

Organ readily admits that the influence of the British drove the Neo-Vedāntists into this fuller appreciation of the world and individuality. Yet Organ believes that Neo-Vedāntism, in its major developments, was not planting foreign European seeds in the soil of Hinduism but was simply nourishing ancient seeds already there. Indeed, Organ goes so far as to claim that Hinduism has grasped the uniqueness of what it means to be human more clearly than any other philosophy or religion. That uniqueness can be seen through the concept of *sādhana.*

For Organ, Hinduism is *sādhana*: it offers a goal to humankind and also the means to achieve that goal. The goal of *sādhana* has been described in many different ways, but the term Organ prefers is *ātmansiddhi*, which means the perfection of the essential nature of man or complete self-realization. Since a number of Hindu sages have attempted to spell out the characteristics of the perfected man, Organ takes pains to point out such models do not represent a wooden legalism but a lofty ideal to be approximated. The Hindu is called not to be a "perfected man" but, says Organ, a "perfecting man."

The perfection toward which the human is to progress is *satchitanānda*, the integration of being, consciousness, and value. Hinduism affirms the objectivity of this triad: beings do exist, consciousness does arise, and the universe is saturated with value. The joyful possibility for man is that he is the place where reality can be aware of itself, where the oneness that underlies all plurality can be discerned, where the values that permeate the universe can be recognized and affirmed. Complete fulfillment of this possibility is the ever relevant but never achieved, distinctively human goal. The joy of this possibility apparently occasions envy among the gods. When a man is born, we are told, the gods are jealous because a god is a god and nothing more, while a man is a man, but more: he is the being who is also a becoming; he is the being who journeys to *satchitanānda.* To cease to become is to cease to be human.

This Hindu quest for perfection presupposes the Hindu appraisal of the human condition as harboring two fundamental possiblities. On the one hand, human beings find themselves in various states of sorrow, suffering, meaninglessness, and frustration. Their situation is one of *dukkha*—misery or *Angst.* Each persons's *dukkha*, however, is the result of *karma* operating through *samsāra. Karma*, simply stated, is the moral law of cause and effect: whatsoever a man sows, that he will also reap. Good deeds produce beatitude. Evil deeds generate misery. The inexorable nature of *karma* is reinforced by the doctrine of *samsāra*, or reincarnation. The consequences of one's deeds will come to pass, if not in this life, then surely in another. One cannot escape from the misery he merits.

On the other hand, the human condition harbors the

possibility of *mokṣa*, the opportunity for liberation from *dukkha*. To attain *mokṣa* (liberation) is to achieve *atmansiddhi* (perfection), and Hinduism spells out the techniques for reaching that goal.

These techniques are the second facet of *sādhana*. In exploring the means of achieving liberation/perfection, Hinduism takes seriously the determinism and freedom which seem to constitute the human condition. With regard to determinism, Hinduism views all human beings as situated in three domains. First, all humans confront four established goals (*chaturvarga*) which must be reached if one is to achieve full self-realization: the goal of hedonic satisfactions (*kāma*), the goal of material possessions and physical comforts (*artha*), the goal of the fulfillment of the duties and obligations pertaining to one's station in life (*dharma*), and the goal of release, liberation, and salvation (*mokṣa*). He who ignores one of these goals does so to the frustration of his own perfection. Second, all humans have four stages (*ashramas*) of life to fulfill: the student (*brahmacharya*), the life of the householder who is married and raises children (*grihasthva*), the preparation for retirement (*vanaprasthya*), and the stage of retirement when one is a student again but now with experience (*sannyasa*). He who ignores one of these stages falls short of human fulfillment by truncating the established nurturing process. Third, all humans have before them four classes (*varnas*): the scholar (*brahmin*), the warrior (*kshatriya*), the merchant (*vaishya*), and the worker (*shudra*). This class structure is an order of quality, *brahmin* being the highest. Each individual is born either into one of these four classes or into the lowly ranks of the outcasts. The position into which one is born is determined by the quality of life one has lived in past incarnations. *Varna*, then, is determined by *karma*: the social class into which one is born is that which one deserves.

Within these predetermined conditions of life—the mandated goals, the prescribed stages of life, and the karmic-established social class—the human being can exercise freedom in pursuing complete self-realization which is liberation/perfection. One's freedom is manifested not only in accepting the fixed conditions of life and thereby aspiring to self-realization but also in the way one pursues that perfection. Apparently taking into consideration the vastly different types of human beings, Hinduism offers four paths (*mārgas*) any one of which, if followed steadfastly, will lead to liberation/perfection. For those persons who are primarily intellectuals, Hinduism offers the path of thought (*jñana mārga*). For those who are worship oriented, there is the path of devotion to the god of one's own choosing (*bhakti mārga*). For those inclined toward benevolent activism, there is the path of good works (*karma mārga*). For those who are fundamentally athletes, there is the path of physical/psychological discipline (*yoga mārga*). In focusing in this way upon human destiny and in being so irenic that room is made for all human types, Hinduism evidences its "human catholicity."

Hinduism sees man as transcending his kinship with plants and animals by virtue of his capacity for self-awareness. Because he is self-aware, man is the being who pursues goals, the being who becomes. The most elevated goal man can seek is his own complete self-realization. Hinduism tells man how to achieve that goal amidst the givens of his situation.

RECURRING THEMES

Although the contributors to this volume write from diverse perspectives and use quite different explanatory categories, nevertheless a number of central themes seem to resonate through their works with respect to both theory and practice.

First, from the theoretical standpoint, to understand humanness adequately involves acknowledging the polar structure of the human condition. The nature of humanness is, it would seem, best described via a dialectical method which depicts truth as the mediation of opposites, as the tension-unity of polarities. Wilson, for example, depicts human creativity as the blend of analytic reason and aesthetic intuition. In a similar fashion, Kreyche explores the problem/mystery polarity of human existence, and Hiltner examines a cluster of paradoxes which must be maintained in dynamic tension in the interest of human wholeness. Gilkey presents the synergism of technological competence and religious commitment as essential for the development of a humanizing society, and Yankelovich argues that the integration of the nature/nurture perspectives is necessary for understanding humanness. Ricoeur displays the ideology/utopia polarity as containing two visions both of which must be kept in tension-unity for the sake of humanizing society. Organ discusses the Hindu view that human reality harbors the twin possibilities of *dukkha* and *mokṣa*, suffering and liberation.

To sacrifice one side of these polarities to the other means to entertain a flawed view of humanness which, in turn, means to court disaster in one's practice. Several centuries ago Pascal seems to have perceived this intimate relationship between theory and practice when he wrote, "Man is neither angel nor brute, and the unfortunate thing is that he who would act the angel acts the brute."[7] Perhaps the ambiguity with which technology has so recently become shrouded—having offered itself as the liberator of humankind from certain external fates and yet in its very triumph exposing humankind to new bondage—has sensitized the contributors to a Pascalian respect for the paradoxical structure of human existence.

A second theoretical theme which pervades the essays is an affirmation of human freedom. All of the writers depict the human being as a process, as the place where polarities meet in ever-renewed unity-tension. Human process, however, differs from that of things by virtue of self-awareness and intentionality: the human is not only aware of himself but he is also aware of future possible selves which he can intend to become. That is to say, human process harbors freedom: the openness to pursue consciously and intentionally a future homeostatic blend of human polarities. Human freedom, however, is not absolute: it too is poised in dialectical tension with those constraints imposed by history and nature.

From the practical point of view, the authors seem to espouse a cautious optimism about humankind's future. To take human freedom seriously, in spite of its boundedness and in spite of the ambiguities to which its exercise in technological development has brought society, is to entertain hope. The contributors do not offer a detailed program for the creation of a humanizing society; such a blueprint would be far too ambitious, and perhaps presumptuous, for a volume such as this. Instead, the writers offer the paradigm of themselves as human beings who have not despaired of the possibility that the creator of an advanced technological society can also be the guardian of humanness.

NOTES

1. Seward Hiltner, *Self-Understanding Through Psychology and Religion* (New York: Charles Scribner's Sons, 1951), p. 6.

2. *Ibid.*, p. 120.

3. *Ibid.*, p. 170.

4. Langdon Gilkey, *Shantung Compound* (New York: Harper & Row, 1966), p. 75.

5. *Ibid.*

6. *Ibid.*, p. 76.

7. Pascal, *Pensées*, No. 358, translated by W. F. Trotter (New York: Random House, 1941), p. 118.

CHAPTER ONE

The Humanness of Physics

R. R. Wilson

Does not it seem incongruous to be discussing the humanness of physics? If one subject would appear to be lacking the quality of humanness, it is physics. This science is characterized by precise measurement and abstruse mathematics; it is rigorous and austere; indeed it is about as objective as a discipline can get. Yet, in spite of a prevalent belief that physics is cold and inhuman, a belief that it has to do only with things, not people, a belief that its Faust-like practitioners blindly and dully follow the rites of scientific method to grind out a plethora of uninteresting facts . . . in spite of all this, I am going to maintain that there is a quality of loveliness in the content and devices of physics, that it is a beautiful creation which has meaning for man's view of himself and his place in the world, and that these qualities of physics can appropriately be discussed under the rubric of humanness.

PHILOSOPHICAL ASPECTS

Although modern physics can be said to have begun essentially with Galileo, I prefer to see its genesis in the atomic physics of Democritus some 2500 years ago. Lucretius gives the most complete description of that ancient atomic theory in his *De Rerum Natura*. This beautiful poem expresses a remarkably up-to-date version of modern ideas about atoms. The Epicureans not only had quite correct notions about atoms, but also about atoms dancing in the vacuum or, as we say now, executing random movements. And if their ideas about force were hazy, to say the least, their idea of the random "swerve" of the atom has inherent in it rudimentary elements of quantum mechanics. The theory is simple and elegant and interesting—

even correct in many respects. But Lucretius's poem is only in part devoted to a technical expression of physics. For the most part, the poem is about the problems of people caused by their superstitious beliefs. Physical occurrences before the atomic theory, no matter how trivial, were explicable almost entirely on the basis of divine intervention. Thus if the wind were to be favorable, then a god or a spirit would have to be propitiated by sacrifice. Agamemnon, setting out for Troy, must cruelly slay his fair daughter Iphigenia. Behind each rock that could stub a toe might dwell a mischievous spirit. These countless little spirits, "always about Man's path and about his bed, mostly hostile by instinct" were only to be pacified by tedious acts of worship. We are told that the lives of the religious Greeks were made miserable by a multitude of gods and spirits, all bullying humans without surcease.

Atomic theory provided an alternate view of how the world worked; if one believed in the existence of atoms and in atomic theory, then that simple all-encompassing theory of every-thing—since everything is made of atoms—made it no longer necessary to believe in spirits as causative agents. Although Lucretius intended his eloquent poem to be a paean to Epicurean materialism, the poem comes out also as a testament to the humanity of Lucretius—it is an example par excellence of the humanness of physics.

Yet another example is an identity crisis for man that was caused by Newton. His simple laws of motion seemed to explain the exact motion of all bodies in the world accurately and universally and very successfully and much more plausibly than did the vague ideas of Democritus. Because of the power and success of Newton's laws, a notion of cause and effect emerged in which it was reasonable to consider the universe to be just one great machine—to be like a great clock that runs all by itself—each motion being caused mechanically as an effect of the previous state of the machine. Once in motion, the positions of every part of that machine would be completely and precisely predetermined for all time. Man, also subject to Newton's law, is just a part of this machine, and hence his every movement also would be equally preordained. In such a view, man is ineluctably trapped by the physics of Newton. He is just a mechanical cog in a mechanical universe. Where is the humanness of this bleak view?

Now this was not the first discussion of the problem of free will or determinism—Greeks and scholastics had enjoyed infinite variation on this theme, but usually involving the nature of God. I think the difference is that however ridiculous or meaningless or oversimplified might be the above caricature of a world based on Newton's law of physics, to this day it still forms the basis of our modern popular materialistic philosophy. Somehow, French Encylopedists and other eighteenth-century philosophers managed correctly to understand the simple physics of Newton and then to draw a plausible inference about what appeared to them to be a completely mechanical and hence materialistic and inhuman world. It seems to me that those quaint eighteenth-century views, based on Newtonian physics and sharpened by nineteenth-century Darwinism, have been pretty much frozen into our literature and have been accepted into general thinking ever since. Perhaps much of the revulsion of some intellectuals to physics stems from an abhorrence of this miserable view of a deterministic materialistic world. If there is any humanness in this bleak picture it is that such a superficial reading of physics should have been taken so seriously, that it should have led to such disastrous thinking, and that it should have instilled such a quality of inhumanness in so many minds. Any deeper consideration of the physics of the problem especially from the point of view of statistical mechanics, or of how matter is actually observed, would have shown the fallacy of drawing any such conclusion about free will from those highly oversimplified eighteenth-century arguments. How much better the lesson would have been had it been one of humility, of a heightened appreciation of the mystery of the universe, had it been one of pride in the human spirit to understand even then so much of that universe.

Unfortunately physics some fifty years ago became so arcane that intellectuals in other disciplines did not notice that a revolutionary discovery in physics, quantum mechanics, made even more obvious the fallacy of drawing conclusions about free will based on Newton's laws. We now know that matter moves as though guided by waves. These waves only determine the probability, the chance, that a particle, or a body, will move to a certain place. Hence the motion of a body moving through space with a precise trajectory according to the certainty of Newton's laws has been replaced by the fuzzy propagation of a wave of chance.

What is more interesting about this from the point-of-view of humanness is that this physics—quantum physics—tells us about the limits of our ability to have knowledge about some conditions of our universe—that there are some questions which we can ask that have no meaning. The theory specifies in a quantitative way just what is certain and what is uncertain. I submit that this precise and surprising information about the limits of man's knowledge is apropos to a quality of humanness. It informs us about ourselves and it leads to a greater appreciation of the mysterious nature of our world.

In the same way, the theory of relativity teaches us about other bizarre but very real phenomena of this world. For example, it is really true that two people can age at a different rate if one person is in motion with respect to the other. Pirandello could have made an even more dramatic study of the nature of reality—a quality of humanness—had he more thoroughly studied the theory of relativity (if he studied it at all!).

Let me turn to a somewhat different aspect of the humanness of physics in man's knowledge of man. Since time began, one could question as did Matthew Arnold about

> The hills where his life rose
> and the sea where it goes -

Originally these questions of from whence do we come and whither are we going were considered to be strictly the province of religion and of myth. What I find utterly astounding now is how modern cosmogony and cosmology—which as a physicist I arrogantly relegate to physics—can inform us about how our world came into being. On the basis of nuclear physics and of various observations, we know that the world began, not with a whimper, but with a bang! Nuclear physics tells us in amazing detail how, from that mighty big explosion about ten billion years ago, energy expanded outward, how the firmament separated from the chaos, how matter in all its forms was produced in known nuclear reactions, and how that matter condensed to form stars and planets. Nuclear physics makes possible the knowledge of the life cycle of stars, about what keeps them hot, how some collapse and some explode, and how some become deep black holes in space.

This partly written "book of genesis" still does not "explain" the "why" of the big bang. However, if we assume in a biblical sense that in the beginning was not the "word," but that instead there was just a tremendous explosion of pure energy from a point in space, a bang, then from that event on our detailed knowledge of nuclear physics has made it possible to give a fuller explanation of "from whence we came" than had ever been given before. The account not only reads like the book of Genesis, it reads like a fascinating detective story. From a very few clues, a more complete *De Rerum Natura* has been deduced. It is a towering intellectual accomplishment, comparable to or even exceeding eighteenth-century poetry or even Renaissance painting. It has added to the measure of man—to his humanness—and his spirit, his human spirit that is, soars out over time and space, and the vehicle of that spirit is nuclear physics!

AESTHETIC ASPECTS

Now let me turn to aesthetic aspects of physics. We physicists are proud of the monumental laws of nature we have been able to formulate. For a physicist, the laws of physics themselves have great beauty. The laws express so much in such elegant form that we can compare them to poetry, and in particular, because of their brevity, compare them to the Japanese seventeen syllable haiku. Unfortunately, just as the haiku is not accessible to most of us because we do not know Japanese, in the same sense, the poems of physics are also not accessible to most of us because we do not understand the language of mathematics.

It is no accident that the laws of physics are beautiful. In groping toward many of the great truths of physics, intuition is an important guide, and beauty is one of the beacons which guide that intuition. Keats tells us:

> Beauty is truth, truth beauty—that is all
> Ye know on earth, and all ye need to know.

But we physicists find it helpful, in arriving at a truth, also to depend upon a few experimental observations. Einstein, however, used a minimum of experimental information in formulating his theory of relativity. He wrote that an aesthetic feeling of "rightness" and a sense of beauty were dominant factors in his thinking. Even when the first experiments of Kaufmann seemed to be in contradiction with his theory, Einstein persisted in belief in his structure—why? Because he felt it was beautiful!

Dirac, the theoretical physicist whose theories first led to the concept of antiparticles, has explained how he gave overriding priority to aesthetics in formulating his theory. He also has pointed out how Schroedinger, one the founding fathers of Wave Mechanics, put forward the "Schroedinger Equation," basic to Wave Mechanics, in spite of apparent contradicting experimental evidence—again because he thought his equation was beautiful. My point here is that aesthetics is an important part of physics.

For example, we physicists like to look for something that is symmetrical in the world. Many of our deepest truths about nature are expressed by a symmetry that we are able to recognize in a property of nature. Thus we like to see that our laws of motion are exactly the same when viewed through a looking glass, that is, when we have interchanged everything on the right side to the left side. That shows a symmetry in space. We might also look for a symmetry in time, by expecting our laws to work as well when time is reversed; for example, Newton's laws are the same for the objects in a movie even when the projector runs backward. Still, symmetry can be boring, so after we have recognized a property of nature that is symmetrical, we are utterly delighted to find any small deviation from the rule.

Tsung Dao Lee who, with Frank Yang, first conceived the idea that the mirror symmetry was occasionally broken, has emphasized the similarity of the slightly-broken symmetry with classical sculpture. He cites many examples of statues that are almost symmetrical but are only lovely because of small departures from exact symmetry. Similarly, this is true of poetry. A too symmetrical poem becomes doggeral; it is subtleness and surprise that characterizes great art—and great physics as well. My point here is that physicists not only use aesthetics to guide them, they also use the language of aesthetics to discuss their subject. Is not all this use of the language of aesthetics just another way of acknowledging a quality of humanness in physics?

In experimental physics too, aesthetics plays a role. For example, the artifacts of physics themselves usually have an innately handsome quality that is quite independent of their function, even though the quality of beauty derives from the function. Thus antique balances, or electroscopes, or magnetometers have now become valuable objects that are collected and exhibited in the salons of the homes of the wealthy. Those objects were designed by scientists who cared about appearance as well as function, or better, who appreciated the relationship between form and function and appearance. Most instruments or machines of physics are designed as a picture is painted; the parts are one with the whole, the whole is one with the parts—all directed toward and expressing the function.

As an accelerator builder, I have found great satisfaction in relating to the men who built cathedrals in the thirteenth century. When Ernest Lawrence built his cyclotrons with a dedicated passion he was not that different from Suger, also with a dedicated passion, building the cathedral St. Denis. The Abbot Suger was expressing a devotion to the church with his exalted structure, a structure that transcended all contemporary knowledge of strength of materials. And Lawrence too expressed, in his fashion, a devotion to the discovery of truth. He too transcended contemporary technology in attaining his dizzying heights of energy. I am sure that both the designers of the cathedrals and the designers of the nuclear accelerators proceeded almost entirely on educated intuition guided largely by aesthetic considerations. This can be seen explicitly in the notebooks of Villard d'Honnecourt, one of the ancient architects: his designs of parts of cathedrals are sometimes mixed up with drawings of the human form.

My own experience has been in designing modern accelerators. These are exceedingly complex machines which are characterized by large mechanical and electrical systems and by complicated forms which are pierced by vacuum pipes, which are immersed in magnetic fields, and in which atoms are joggled by electric fields, and jiggled by electronic devices. Informing and controlling this complex is a nervous system that consists of a ganglia of microprocessors and a gaggle of computers. Now to understand each complicated component and its relationship to the whole would go well beyond my own technical knowledge. So how do I go about designing? Well, I find out a little here, by a simple calculation, and a little there, by another calculation, about the parts of the most important technical components. I then draw those parts of the design on paper. After that, I just freely and intuitively draw in neatly appearing smooth lines, lines which cover my ignorance of detail. I keep drawing, correcting here and there by calculations until the accelerator appears that it might work. Mostly I know it might work because it looks and feels right— not because of any long and detailed chain of calculations, which after all probably could take forever. It is when the parts and forms have essentially the same relationship that the parts of a sculpture should have to the whole that I am satisfied by the design. Of course, building, even designing, a large accelerator is a complex team activity—just as it was for the cathedral. But if the conceptual design is aesthetically right, then one can depend on the members of a team to appreciate and to understand and to respect that aesthetic form in their own creative contributions. An accelerator that is "understood" works. What I am trying to express by a certain amount of exaggeration is that most of the effort of design is intuitive, that aesthetics is indeed a valuable and necessary guide in any design process, that these very human qualities are an important part of physics and give to physics a quality of humanness.

CREATIVE ASPECTS

Physics is not developed by mechanically following an arbitrary set of rules that lead automatically to new knowledge. For example, just blindly following the so-called scientific method will get you nowhere. Now, the physics that ultimately appears in textbooks is usually expressed in an elegant, but nearly incomprehensible, mathematical form. It is very different, however, during the process of discovery. Not too much is known about how the mind of a physicist comes to produce new physics, just as not much is known about how the mind of the artist comes to produce new works of art. One thing is clear, it is that there is much in common between what the creative artist does and what the scientist does. Scientist and artist even use similar words in describing their creative moments. They both care deeply about what it is they do and they tend to be single minded in going about doing it. They both usually pass through a morose period when their contemplation is deep, and mostly at a subconscious level. Finally, what they do create just comes bubbling up from the subconscious to the conscious mind. Inevitably this inspirational act is accompanied by a period of euphoria, an ecstasy which has been documented time and again for scientist and artist alike.

What I have been trying to maintain is that those qualities that we identify with humanness are as active and important in physical research as they are in painting or poetry. The artist and the scientist appear to take on very similar roles at the level of creating knowledge. Even their personalities are similar—and they both tend toward the idiosyncratic. Take Einstein and

Picasso: they both dressed in sweat shirts, neither wore socks, and both were very strong, individualistic characters.

If the act of creativity in science and art is similar, I would like to go now farther and to question if there is not a similarity, as creations, in the content of art and of science. I want to ask if both are not equally creations of the human mind? And I do not mean by this to deny, in any way, the independent existence of that mysterious but objective reality "out there." However, there are many ways that we could come to perceive that reality, and the particular way that has evolved and the particular corner of reality that has been explored might both be considered to be arbitrary creations of the mind of man.

The fundamental concepts themselves, such as distance, or mass or time, are idealizations of nature, and they had to be invented by man. Perhaps it might be more accurate to say that such concepts had to evolve with our language, but after all, language itself is a human creation. For example, the concept of time which has evolved for peoples with different languages appears to be quite different; we all experience the remarkable difference between the passing of psychological time and of clock time. The art of science has been to create concepts, and then to create theories using those concepts that correspond to reality.

If it is at all plausible that the concepts of physics themselves have at least in some degree been arbitrary creations of man, rather than being "revealed truth," or being something that was just inherently obvious to anyone, then it might also be plausible to go on and to inquire if all knowledge of physics has not in large measure been created by the mind of man—just as Dante created his many-leveled universe of Paradise and Inferno. The difference is that physics must constantly, step-by-step, be consistent with the world of reality as determined by experiment. And of course the objects of our reality have a continuing reality which is largely independent of how we perceive them.

What I am saying is in a sense just an exaggerated form of the three worlds of Karl Popper. His first world has to do with things, his second world is that of subjective experience of thinking. His third world consists of statements and theories and of mental constructs, and is one which I am sure he would agree is a creation (in a evolutionary sense) of the mind of man.

But how can I maintain my belief in an objective reality and at the same time suggest that the world we know is even some part, a creation of man? To try to clarify that dichotomy, let me compare the physicist creating knowledge to an artist creating a painting.

The artist starts with a blank canvas, some paints, and the tools for applying them. He makes a few initial strokes that pretty much determine the final nature of the picture. These initial strokes might be analogous to the fundamental concepts with which the physicist starts. The ultimate picture may correspond to one of our theories. The artist can start again with exactly the same amount and kind of paint, and then use it all up to complete a new picture. From a chemical and a physical point of view, the two pictures he has produced might be judged to be very nearly identical—but the meaning of the paintings to someone observing them can be utterly different. Indeed, any number of such "identical" pictures could be made starting with exactly the same ingredients, but all could turn out to have utterly different meanings and all be original creations.

My analogy, then, is that the canvas and the paint, which would be physically and chemically the same for all the paintings in this example, and which are quite independent of the painter, correspond to his physical reality. Without violating that reality, the painting is made, and it will come out differently each time; each is a creation even though the painter does not create the paint or the canvas. So might it be for the physicist. His reality is more mysterious, and more circumscribed, but it is plausible that were man to start again from his beginning, then the science which would evolve and hence the world he would perceive might appear different to him than it does to us. In that sense the world we perceive is a creation of man; in that sense, it is a manifestation of the humanness of physics, and in that sense, we do not have new worlds to conquer, or even new worlds to discover. We have new worlds to create.

If man does create in some sense his world, then he has a responsibility for that world—a responsibility for the success or failure of that world. To my mind one of the greatest failures, in spite of what I have been saying, *is* the inhumanness of the world of physics. Physicists, not unnaturally, have taken the business of physicists to be the creation of physics. The business of physicists should also be concerned with the poetry and drama of physics, with the beauty of its form and function.

The business of physics should also be the business of humanists and artists.

I take the optimistic view of man. He will survive the inhumanities of his technological culture, and I believe that his culture will evolve to a new and richer level—that he will create a new world. But I agree with A. N. Whitehead that "unless we can make man, his culture and his ideals, of central importance to the physical scientist *in his* own work, we are in serious danger of sinking to the level of technologically skillful barbarians." Perhaps in the humanness of science and art, the humanists, the artists, and the scientists can find a unity in their cultures.

CHAPTER TWO

The Meaning of Humanness: A Philosophical Perspective

Gerald F. Kreyche

It is itself a sign of humanness that in our era of rapid change, we can all take time out to discuss the meaning of humanness. It is good to talk about humanness, for philosophers too seldom venture beyond their own collective history. This collective history, akin to the laboratory world of the scientist, is quite other than the ordinary world in which they live, and move, and have their being. I am personally sympathetic to the charge made by a prominent American philosopher some five years ago that while "philosophy is not an intellectual wasteland, it is dreary."[1] Or as the French leftist, Paul Nizan, put it, "Philosophy is not dead; but it needs to be killed!"

In opening wide its doors to technicians, philosophy now runs the danger of being nibbled to death by the minnows of the world. Despite current trends in other disciplines, the fact is that "a philosopher is inevitably a generalist. In an age of specialization a high price must be paid by the generalist for unavoidably being somewhat superficial and out of touch with the latest developments."[2] But human nature is not one of the latest developments, and I hope that this paper will remove some of the sting from the indictment given earlier.

Our task involves both a quest and a question. It is a quest, for despite differences, we share a commonality with the past and must re-examine that past as one step toward discovering who we are; that is to say, not only as individuals but as humans. Indeed, this is part of the problem facing us today, namely, upholding a common human nature, all the while recognizing the uniqueness of each person.

It is a question, for we now philosophize out of a different *Umwelt*, a different perspective and a different life-style than did our forerunners. Our world views and horizons have shifted perceptibly, and often we find that what we once regarded as human we now regard as inhuman and conversely.

Aristotle's description of man in the terms of *zōon logon echon* seems unreal to many who feel a closer affinity with Kazantzakis' Zorba, the absurd and sensuous man.[3] This same span of history finds a different appreciation of woman, formerly being designated a chattel, now referred to as a Ms. (Or should we say in those states refusing to ratify the Equal Rights Amendment, a near Ms!) It is a span which once enshrined a universal human nature in a Platonic world of ideas but today, with Sartrean denial, claims for man only a common condition out of which he creates himself and becomes the ensemble of his actions.

In short, the quest and the question of humanness has moved from that of "What is Man?" to that of "Who am I?" It has moved from the plane of individual to that of person. The horizon against which this inquiry is raised has changed from a spectator world to a participatory one. Presently (and primarily in the United States), it is a world which extols the individual over the institution, the amateur over the professional, and freedom over knowledge. It has come to be a world largely of our own making, this through the resources of science and technology and the insights of phenomenology.

Having made this prologue, let me now present the order of future discussion in this paper. First, I would like to paint with a broad brush the characteristics of our time which have a bearing on our appreciation of humanness. Next, I will consider the problem of humanness under the aspect of what it is to be a person. Lastly, in order to understand person as concretely as possible, I will offer a philosophical reflection—a meditation, if you will, hopefully exemplifying the manner and depth of insight all of us can have regarding our humanness if we but take the time to do so.

OUR AGE

Our times reflect a period of questioning, soul searching and identity seeking. Jacques Maritain and Karl Stern see this in part as the result of four insults cast our way over a period of several centuries. These may be termed the astronomical, biological, economic and psychological insults and were hurled by Copernicus, Darwin, Marx and Freud respectively. The first displaced man's world from the center of the universe. The second revised man's views of understanding himself more as a risen animal than a fallen angel. The third gave a new explanation of human progress and shaping of the masses, and the last grounded man's actions on the hidden foundations of the Id and the massive unconsciousness rather than on free acts.

No wonder that our times give birth to a collage of personalizing and depersonalizing factors. It is unfair to blame the latter exclusively upon technology and to wring one's hands about the future à la Guardini or Ellul.[4] It is my unromantic belief that technology has wrought far more good than evil and that what is needed in a world of 3.8 billion people is not less technology, but more technology; that is, unless we wish to go back to the human life lived only by a few wellborn, with the rest eking out but a subhuman existence. It is not technology but in its misapplication that depersonalization occurs. For example, it is technology, in its broader understanding, that has brought us mass production and the discount store which moves enormous volumes of merchandise at minimal profits. It is the discount store—the K-Mart, the Goldblatt's, the A & P—which have enabled the many poor to dress tastefully and in respectable fashion and to eat moderately well, thereby enabling them to walk inconspicuously side by side with the customer from Nieman-Marcus. The scholarly paper has yet to be written which shows the evident contribution of the discount store to the *bonum commune* of a nation. This unheralded breakthrough is no less a revolution than that ushered in when the producer became the consumer. But enough of this apology, for I want only to describe factually our situation today.

Note the following advances that have been made regarding the concrete recognition of humanness. These are largely through institutions deferring to the freedom of persons. Nurses may wear pantsuits in hospitals, patients their own pajamas, and the curtains of the room need not be gauzelike in color and texture but highly and cheerfully decorated. Our society has encouraged various liberation movements among women, blacks, students, prisoners, children, gays, golden agers, etc. Foodstuffs are dated, and the pluralism of ethnic lifestyles is not only tolerated but encouraged and even imitated.

Any one of these may seem inconsequential, but cumulatively they represent a symbol of hope and trust in the ability of human nature to perfect itself, not in isolation but in community with others. Yet this community embodies a different understanding of man than that in the Greek sense of *zōon politikon*. While man is a political animal, this no longer means that he relates to the state as an individual to the species in the animal kingdom. Happily, we have done with the pyramids and cathedral builders whose vision of the individual man was akin to their vision of the queen bee or the egg-laying salmon. Such a vision apprehended the individual's worth only in the contribution one made to the whole; in short, the individual's worth lay exclusively in his jejune functionality. This was the day in which man was modeled after the state (cf. Plato's analysis of virtue in man through an appreciation first of virtue in the State).

Perhaps we now have moved to the opposite extreme, for in our crisis-ridden age, we look askance at the institution. It seems to many to be unreal. Hence, they revolt when asked to serve it without question, even though it offers to accept total responsibility for those who do so—a clear instance of Sartre's analysis of *mauvaise foi*. The fact is that although institutions of various stripes are seen as having diminishing worth, we need them desperately in order to secure our human rights. I believe it was Aristotle who said that whoever has no need of society is either a beast or a god. (This institutional nominalism, which exhibits Rousseauian roots, is as great a danger today as a spreading bureaucracy. Unhappily, the abuses of each provides grist for the other's mill.)

On the other side of the coin, we have often been identified with our social security numbers; the privacy of our inner and outer being has been bombarded by immense forces of propaganda which, invading the subliminal, conditions and tells us to do this, buy that, go there, etc.[5] Many of us are concerned more with our image than our real selves and portray only the functional man who is identified with his operations. Even our clothes show how caught up we are in the spirit of advertising, for clothes now have become costumes to be worn in our role-playing activities. Surely it is time to question values when in fifty years, one hundred million people have been killed by people. Auto accidents in the U.S. alone yearly total the number of American soldiers killed in the recent Vietnam conflict.

More subtly, perhaps, we witness a reversal of human values in our readiness to accept a comparison of the eye to a camera, the brain to a computer, and the body to a machine. Strangely, man's creation has become the prototype and paradigm of man himself. It is wonderfully ironic that we marvel how man can fly so complicated a thing as a modern jet airplane. Marx was right, not so much in his example, but in seeing how we have so alienated ourselves that we now stand in awe of our own creation, and forgetting it is ours, bow down and worship it. If I be permitted a neologism, then I can say that we are no longer anthropomorphic so much as we have become "mechano-morphic."

The times are indeed strange and offer us many temptations to sell our humanness for the tinsel of efficiency. Such depersonalization offers us bribes of immediate favor which in the long run serve only to corrupt us. But in spite of dial-a-prayer, bingo and bumper stickers, we have not lost completely the feel for humanness which is exemplified by the embarrassment still experienced at calling to an elevator to wait—then realizing it was an automatic. And perhaps there will even be some spin-off benefits from our obsession with a costing mentality such as that proposed not long ago in a congressional bill for a "person-depletion" allowance!

These then are the times in which we live, and if we are to discover the meaning of humanness in them, we must look for this meaning not in the abstract form of humanness (since there can be no *a priori* analysis of it) but in the concrete experience of person, for person is what is most concrete in nature.

TOWARD AN UNDERSTANDING OF PERSON

Person relates all to itself and itself to all. As Maritain puts it, "person is what cannot be indifferent to things."[6] Person is what is centermost, then. Even in the contemporary philosophy of religion, person has become the creator-bestower of value. Louis Dupré expresses it well:

> To the modern believer something is sacred mostly because he *holds* it to be so by internal conviction and free decision, not because he *undergoes* its sacred impact. Thus he bestowes the quality of the sacred on realities which did not possess it before and withholds it from objects which his tradition has always regarded as sacred.[7]

The centrality of person, having become so concrete, has forced abstract truth itself to undergo a change of status.

Increasingly, it is not viewed as meaningful except when concretized in a value. Only thus can the abstract universal of truth, i.e., by being brought *up* to the level of a particular become a valuable, and so that for which a person might live or die. Put more technically, abstract truth must undergo a kind of reverse transcendence in order to have existential import for the life of the person. The Danish nineteenth-century religious figure, Søren Kierkegaard, saw this when he proclaimed, "Truth is a snare; you cannot have it without being caught. You cannot have the truth in such a way that you catch it, but only in such a way that it catches you."[8]

Person, then, is pivotal in the world of humanness. Like love, person is a totality, a world within a world, rather than apart from a world. Person is one who is on the threshold, who lives on the border, as the psychiatrist-philosopher Karl Jaspers puts it, in a *Grenzsituation*. That border manifests itself in many ways: the border of the material and the spiritual, the border of certitude and skepticism, the border of faith and knowledge. This borderline situationality, exemplifying the humanness of man, can be seen in the following: in faith, man experiences the tension of opposites. If he is certain he has faith, he can be certain he does not have it. He is fated to occupy the in-between-land bounded by presumption and despair.

Or again, the extent of his freedom is never fully known to him in this or that single act. There, reflection shows him three perspectives, all vying for acceptance, all dialectically different. These obtain when he considers separately the extent of his freedom *before* he acts, *while* he acts, or *after* he has acted. To take one perspective as the only and total truth—to choose sides, so to speak—always results in a kind of fanaticism.[9] The admission of freedom is undeniable. Although all arguments may be against freedom, William James tells us, all experience is for it.

Nonetheless, man's freedom is always a "becoming" thing, and historically he has experienced precariousness in having to make harried efforts to preserve his freedom when squeezed between two forces which would deny it. These are the forces of worldly determinism (whether marshalled by a Democritus or a Skinner), and that of a misunderstood notion of Divine Providence and predestination (as may possibly be rooted in a St. Augustine).

Man's humanness as person is still on the endangered species list because of the stubborn quest for the status of scientific recognition by disciplines formerly humanistic in outlook. If there is a threat to man's humanness, it may well be more from within academe than from without. The methodologies of psychology and sociology, for example, today increasingly attempt to imitate the ways of physics and chemistry, the so-called "hard sciences." Thereby these "soft disciplines" implicitly wish to view man as a "thing among things," a denizen of the lesser world.[10]

In light of the above, we can appreciate Ortega y Gasset's characterization of the intellectual as hero. The intellectual is forced to act, to tread firmly the grounds of uncertainty, while others only stand still trembling. No wonder Ortega (quoting Herbart) can write, "Every good beginner is a skeptic, but every skeptic is only a beginner."[11]

THE DRIVE OF SPIRIT

At the root of person lies the reality of spirit. Whether known as *psyche, anima, Geist, ruach* or whatever, it is the touchstone of all that is real. Yet it is impossible to touch or define. It is movement, life, generally progress and all in all, is alive and well. Perhaps we can appreciate its reality descriptively and through example, as well as running through the visions of great philosophers who have discovered it.

Spirit has been with the world as long as person. Sometimes it takes on an unsure and wispy character; at other times its force is evident and overpowering. St. Augustine detected it in his *City of God* and gave us a mighty theology of history. Both Hegel and Marx saw it, as giving unmistakable direction for their own respective philosophies of history. Henri Bergson called it the *élan vital* and Teilhard de Chardin envisioned it as the movement from the alpha to the omega point.

Spirit has welded individuals into groups (esprit de corps); groups into a nation (The Spirit of '76) and it has propelled a nation to goals bigger than life, thereby producing actions both famous and infamous (The Spirit of Manifest Destiny). It has captivated an age (The Spirit of Romanticism) and has directed modes of thought (The Spirit of Technology).

At times, spirit beckons falsely, or as Bergson puts it, moves into blind channels from which it must retreat in order to go forward once again. Its false beckonings in certain forms of romanticism can be found in the nostalgia of our times, urging us to return to the "natural life" of the Indian in the wilderness. It beckons falsely in those enamored with an excess of the spirit of technology, which excess would have reason "manipulate the entire earth and all living things within it."[12] The latter has many forms such as bio-engineering, e.g., cloning, computer dating, and in general the hucksterism in academe of management by objectives and competency based education. Because of this false beckoning, at Watergate we nearly equated a moral fault with a kind of error in strategy or game plan, a type of mathematical mistake. Because of it, we have tended to misread tragedy as a kind of error rather than an exemplification of the spirit of evil which also abounds.

PERSON AS SUBJECT AND SUBJECTIVITY

But if the principle of person rests in spirit, the fulfillment of spirit is realized in person. With person, the subject and subjectivity were introduced to philosophy. That is why the distinction employed by August Brunner between *erkennen* (to know or to understand) and *erklaren* (to clarify or to explain) is so important. We have a faulty picture of person if our knowledge of him is garnered only by *erklaren*. This is the kind of knowledge science produces and is similar to the calculative reasoning *(berechnendes Denken)* spoken of by Heidegger. Persons are primarily to be known in the special way of *erkennen,* and to try to explain the latter by means of the former is to be guilty of the same kind of error as those who tried to define the good in terms other than the good.[13]

Subjectivity has entered the portals of philosophy then, much to the protests of those who would continue philosophy's pretentiousness of total objectivity. Indeed, this perspectivalism shown by subjectivity has always been part of philosophy as any cursory study shows. The conditioning elements of a given culture have always encouraged philosophical outlooks, selection of problems and solutions to those problems.[14]

Although spirit is a principle, it is person who is the ground of value and meaning. As indicated earlier, phenomenologically and technologically more and more, the world appears as of our own making. We understand it only in relationship to us and conversely, not unlike the point made by Hegel in his famous example of the master-slave relationship.

Accordingly, person's relationship to the world cannot be that of an onlooker but rather of one who is caught up and involved in it. This means there is no escape from the human perspective and condition. That is why a theology will always be in part an anthropology at the same time, with the reverse being true as well. In such a perspective, person is seen as an absolute, a center, a focus of all that is sacred.

Let us now see more exactly how personalism affects a philosophical understanding of humanness. (In what follows, my general indebtedness to Gabriel Marcel is so patent as not to need acknowledgment in the specifics.)

PERSON AND MEANING

Marcel points out that the greatest error of our time is the identification of *being* and *having*, thinking that a person *is* someone only so long as he *has* something. Similarly, as we have confused these two orders, we have confused what belongs to them as well. Examples are wonder and curiosity, wisdom and science, friend and acquaintance, etc. Wonder, wisdom, friend all belong to *being*; curiosity, science, acquaintance to *having*.

Although these constitute two orders, at times they do approach each other as the following example shows. I not only *have* a body, but in a certain sense may say that I *am* my body. My body is not simply an instrument for some "ghost in the machine," but it is really me. As a friend might exclaim, upon grabbing my arm after not seeing me for many years, "It's *really* you!" The person of man, then, is his whole being and can be called an incarnate spirit. As such, he can dignify and spiritualize. He can confer value on what he touches so long as he respects its own nature as well as that of the other persons. Marx appreciated this human conferral of value upon work by man. But it is no more true there than it is in art or play or religion. Put in more classical terms, the effect in some way will always mirror the cause, since the effect itself is really the prolongation of the agent in the patient.

Unless we misconstrue value, restricting it to "pragmatic uses," we see the world of value is constituted largely by man as person. If a thing is loved because it is valuable, i.e., has a pragmatic use, we have not risen above the level of *having* to that of *being*. We reach the latter only when we see that a thing is valuable because it is loved. Only in the latter case have we assimilated ourselves to it, allowing it to share in our being.

Some examples will clarify this point. The meaning and value of something as prosaic as vacation slides or a family album can be realized only by those who are related to it, *because the slides and album are related to them*. To the outsider, they provoke sheer boredom. The ring in the store window and that same ring on the finger of the young lady having had it given to her with a promise by her beau are materially the same. Yet the meaning and value are wholly different, for the ring now symbolizes and reveals a human relationship. In this sense it can be said that humanness means the worth of the gift lies principally in the spirit with which it is given.

A PHILOSOPHICAL REFLECTION

The centrality of person in reality and what this reveals about reality itself may best be seen under some typical conditions in which we confront our humanness in situations such as the experience of grief, hope, faith and friendship. A meditation on mourning, for example, on grief (as opposed to hysteria) on the occasion of the death of a loved one, argues to the continued presence of the other. Just as it is the "we that creates the I," the insight into myself on this occasion evidences the continued presence of the other, albeit under a different mode. Were this not true, this deepest of all human emotions would betray itself ontologically.[15]

Person, then, to some extent, represents the unknown and will always remain in part the unknowable. This in itself constitutes a block to the spirit of aggrandizement engaged in by all those who suffer to wear the "armor of intelligibility." As a partial unknowable, person provides the ground for hope as Pannenberg observes.[16] With Marcel, we can see the meaning which person gives to hope, as distinguished from wish. Wish signifies the unexpectable and unreal; hope indicates what is expected but cannot be fully known. Thus, we say to a friend who is departing, "I *hope* to see you soon!" In the proper sense, hope itself is a projection of faith, an extension of faith, which in turn is an extension of the person. It involves active willing more than intellectual insight. Often we can learn much about it through contrast with despair—a condition in which will becomes inactive.

The faith referred to above is itself grounded in person. Says Marcel, "To believe that . . . I must first believe in." To believe that a thing is so, I must believe in the person who has related the faith fact to me. This means that faith primarily deals not with objects, but with subjects. It reveals the subject both of the believer and of the one in whom he believes. (This is but one important difference between religion, such as Christianity, and ideology, such as Marxism. It is also a distinguishing mark of the *personalism* of Christianity from the *individualism* of Ayn Rand, the *humanism* of Jean-Paul Sartre, or the *collectivism* of world communism.)

Faith will not be so much knowledge-yielding as it will be love-promoting. Yet the latter also helps give knowledge—the ancients called it connatural knowledge. This is also why the first requirement for the lover is that he be full of faith, i.e., faith-ful. Faith, after all, is the promise of our totality, not of a part but of our entire being. Unfortunately, however, just as to the nonlover love makes no sense and is mistakenly called blind, when in fact it opens up a whole new horizon; so to the unbeliever faith makes no sense, when in fact it too opens up a whole new vision, not of seeing a new world but of seeing the world in a new way. (The approach is clearly sympathetic to Augustine, who never tires of quoting Isaias, "Unless you believe, you shall not understand.")

From the perspective of the person, then, being is relational. Even though truth may be infinite and eternal, the person's participation in it is necessarily finite, historical and relative. This relational aspect is nowhere clearer than with the lover. For the person in love, *to be* is *to be with*. His being lacks meaning and value precisely to the degree and kind of separation from the beloved.

Now, the "beloved" is just another name for friend. Here again we see that the meaning and value of person is corelational. Friendship is always the Waterloo for a certain kind of empiricistic mind. It constitutes a scandal, for it is at once the most empirical of all experiences yet cannot be approached or understood by empirical means. It is the friend who paradoxically reveals me to myself, exposing all the good present within me, but which up until now had only been latent. It is my friend who reveals my humanness by bringing to light the transcendent principle within myself through exposing my capacity for sacrifice. This perception of friendship is clearly at odds with the "Dale Carnegie" colonization of the other which we are urged to follow if we wish to climb the ladder of business success. Here one might be best advised to treat friends equally. In personalism, however, that would be the height of injustice; one must treat friends not equally, but differently, for they are persons. Should we see our fellow man only as a "what" instead of as a "who," we see him only as a universal, as a "for me," as a tool, an object, a thing among things. We only grasp him as an "it." Ultimately, this is the ground for slavery and for paternalism, whether parental, civil or ecclesiastical.

To appreciate the commonality of humanness, we must appreciate the uniqueness of the person. In a certain sense, persons are God's "designer collections," each an original—no copies. We unconsciously demean that uniqueness with our propensity to classify and categorize everything and everyone. For example, we see a flower or a stranger and then proclaim victoriously, "That is a rose," or "He is a Black." This gives the impression that we have said something essential about the thing or person, when in fact we have done the opposite.[17]

CONCLUSION

To search for what humanness is, then, is to discover that basically it consists in our innate *ability to become more than what we are*. Neither animal, angel nor God can claim this. In this sense, Heidegger says that humanness is rather constituted by one's possibilities than by one's actualities. Of course, it is not only philosophers who appreciate this fact. It can be found in great and lesser literature, written for all ages. It is found in John Donne's *Devotions*, Saint Exupéry's *The Little Prince* and more recently in a beautiful book by Trina Paulus titled, *Hope for the Flowers*. There a young caterpillar (let us call him a student) is listening to an old caterpiller (let us call him a teacher):

> Said the old caterpillar, "A butterfly is what you are meant to become."
>
> Replied the youngster, "How can I believe there's a butterfly inside you or me when all I see is a fuzzy worm?"
>
> Biding his time, the old caterpillar shows the youngster how to make a cocoon from his own innards. With this humble beginning the young caterpillar suddenly has an insight.
>
> "If I have inside me the stuff to make cocoons, maybe the stuff of butterflies is there, too."[18]

And so it is with human nature.

NOTES

1.　Abraham Kaplan, "The Travesty of the Philosophers," *Change in Higher Education* (January-February 1970): 12.

2.　*Ibid.*, p. 19.

3.　One must stress the affinity for Zorba's absurdity versus the repugnance one feels about Albert Camus' absurd man, whom Camus describes as follows: "The absurd man thus catches sight of a burning and frigid, transparent and limited universe in which nothing is possible and everything is given, and beyond which all is collapse and nothingness. He can then decide to accept such a universe and draw from it his strength, his refusal to hope, and the unyielding evidence of a life without consolation." *The Myth of Sisyphus*, trans. Justin O'Brien (New York: Vantage Press, 1955), p. 44.

4. See Romano Guardini, *Man in the Modern World*, trans. Joseph Theman and Herbert Berke (New York: Sheed and Ward, 1956); and Jacques Ellul, *The Technological Society*, trans. John Wilkinson (New York: Knopf, 1964).

5. This privacy is grounded on our right to limit access to ourselves. Odors, smoking, the raucous noise of blaring rock music from the omnipresent transistor radio violate our privacy as surely as covert wiretapping. Privacy really has to do with the quality of our space. See Roland Garrett, "The Nature of Privacy," *Philosophy Today* 18 (1974):263-84

6. See Jacques Maritain, *The Person and the Common Good*, trans. John J. Fitzgerald (New York: C. Scribner's Sons, 1947).

7. Louis Dupré, "Has the Secularist Crisis Come to an End?" *Listening* 9, no. 3 (Autumn 1974):17. In a way, Kant himself was touched by this outlook on religion. He thought that an established church tended to stand in the way of the relationship of the individual to his God. It tended to block out an authentic morality by pushing us instead to a moral heteronomy.

8. Søren Kierkegaard, *The Last Years: Journals 1853-55*, ed. and trans. Ronald Gregor Smith (New York: Harper and Row, 1955), p. 133.

9. Not long ago, the Christian moral theologian took the perspective of freedom in prospect, i.e., of freedom *before* the act, with the consequence that he saw mortal sins falling out each time one shook one's pantscuffs. On the other hand, the Freudian saw the act in retrospect, i.e., *after* the act, and explained freedom away on the basis of his rationalistic outlook.

10. One should make no mistake about the thrust of contemporary psychology. The scientific community within it has bid adieu to what many considered the embarrassment of a Carl Rogers, Abraham Maslow, Rollo May, and others in the "existential tradition." B. F. Skinner is their paradigm. In this sense, Carl Rogers is correct when he claims that "the basic difference between a behavioristic and a humanistic approach to human beings is a philosophical choice." Carl Rogers, "In Retrospect," *The American Psychologist* (February 1974):118.
This is ironic, as Ortega y Gasset observed, commenting on the crises of principles in physics in the nineteenth century. At that time, physics wished to be metaphysics, and metaphysics to be physics. A sign that both grew to maturity was their later willingness to accept themselves for what they were and could do. See *What Is Philosophy?* trans. Mildred Adams (New York: W. W. Norton & Co., 1960), p. 55.

11. *Ibid.*, p. 172.

12. Gabriel Marcel, *Searchings* (New York: Newman Press, 1967), p. 43.

13. *Erkennen* implies understanding a person from within. To do so we must be free to step out of our world and put ourselves in "the shoes of the other." It also requires that the other so permit us. Personal knowledge really comes first in our world; it is only later that we are encouraged to discard it and substitute instead the scientific *erklaren*. Cf. examples in children and primitives who personalize everything.

14. See Gerald F. Kreyche, "Ethnicity and Philosophers," *Intellect* 104 (1975): 123-25. This article touches on the importance of the psychology and sociology of philosophy.

15. This is made clearer by an example of the Viennese logotherapist,

Viktor Frankl. He stresses that the greatest argument for the existence of water is the empirical fact that persons get thirsty. So, too, the argument for God's existence. One's first response to this stance is to claim its explanation is rooted in the psychology of man. Frankl agrees but points out that the psychological itself is grounded in something more fundamental, namely, the ontological order.

16.   According to Pannenberg, there can be hope only when that in which we hope is partly unknown and partly uncontrollable by us. This involves a risk commitment, hence we have a tendency to eliminate the unknown by accepting only the known. In this way we satisfy our urge to control all. Thus for the Greeks, hope was an evil from Pandora's box, whereas for the Christian it is a fundamental virtue. See Wolfhart Pannenberg, *What Is Man?* trans. Duane A. Priebe (Philadelphia: Fortress Press, 1970).

17.   Says Marcel, "To classify the flower is not an exhaustive answer; in fact it is no answer at all; it is even an evasion. By that I mean that it disregards the singularity of this particular flower. What has actually happened is as though my question had been interpreted as follows—'To what thing other than itself can this flower be reduced?' " Gabriel Marcel, *Faith and Reality*, vol. 2 of *The Mystery of Being*, trans. René Hague (Chicago: Henry Regnery Company, 1951), p. 14.

18.   Trina Paulus, *Hope for the Flowers* (New York: Newman Press, 1972), pp. 71-72.

CHAPTER THREE

Theological Perspectives on Humanness

Seward Hiltner

Reference to humanness—or to human nature, humanity, or humanist—as something positive and affirmative is relatively new in most Western Christian theology. One reason for this attitude in the immediate past is the use of the term "humanist" by persons and groups who wanted to show, along with their positive declarations, that they did not believe in God in the sense of any of the Western theological traditions. The deeper reasons for hesitation in affirming and valuing humanness, however, go much further back in history. They focus around a conviction that mankind or humankind, even though not causing all visible evils (like earthquakes or hurricanes), nevertheless has complicity in, and often accountability for, most of the serious ills that afflict human beings. Unless persons or societies are reconstructed, (or saved, redeemed, reconciled, or made new), the danger possible in appearing to affirm humanity in its *status quo* form appeared so great that the usual course followed was to be negative about humanness, both of persons and in general, in its manifestly existing condition.

The reluctance to affirm human nature or humanness on the part of most Western theology is seriously misunderstood, however, if it is interpreted out of context and only at the surface level. From the beginning of ancient Jewish reflections, the essence of which was carried over into Christianity, it was believed that no adequate statement could be made about humankind in its actual relational situation except one of paradox, that is, an apparent contradiction and hence continuing tension, but not, in some deeper sense, an actual contradiction. The most obvious statement of the paradox appeared, of course, in the form that we now call myth, in the two stories of creation that open the book of Genesis. Mankind is shown there as having been created originally good or righteous. Even though the people of Eden are represented as quickly falling from the original condition, the story is clear that what they are after the fall is not to be understood as most basic. The emphasis is on "isness" rather than "oughtness," on the most fundamental view of created human nature rather than on a humanly constructed view of an ideal nature. In a sense that could be articulated only through myth, specifically through the past time behind time, or what Mircea Eliade calls *"illud tempus,"* the story affirmed that they had experienced a deeper reality of their own nature than the existential situation could disclose even with the help of the best human ideal constructions or projections.[1] They had been there. Things now might indeed be bad all over. But that badness is not the whole truth, or the most basic truth, about humankind.

The Eden stories are complex as are all deep-reaching myths; and it is easy to show, for instance, that any return of allegedly fallen humanity to the original condition described in Eden would be a retrogression, what Kierkegaard called a state of dreaming innocence, a kind of childlike pre-humanness.[2] Critiques of that kind are important to guard against possible attempts to attain the non-tension state of Eden with little work or responsibility required, pleasant walks with God and other humans, and no danger so long as the advice of the serpent is not taken. But such analyses miss the mark if they overlook the deeper belief of the myth, namely, that in spite of how bad things are in both the external and internal worlds of mankind, the more basic created nature of man is not a simple reflection of the existential situation. Thus humankind is indeed not only full of evil but also of sin, the latter term implying complicity in producing the predicament and also, therefore, provided the proper help can be found, some chance of reconstructing the existing situation.[3] If there should be a denial that the existing situation is really bad, there would be no openness to whatever could be the source of reconstruction. But if mankind did not have some quality of "isness" beneath the existing situation, then no source whatever could find a bridgehead to produce or help produce the reconstruction. Hence the basic paradox of humankind as actually being at the deepest level something in contradiction to the obvious realities of both external and internal human existence. Both declarations must be made. The tension must be maintained.

ANIMAL AND IMAGE OF GOD

Another perspective upon the Western theological understanding of humanness appears in the paradox, even more subtle than that of original righteousness and the fall into existence through sin, of human beings as animals, or animated bodies, and also as being made in the image of God the Creator. The greater subtlety emerges from the conviction that, in mankind's most basic "isness," there is no contradiction between living as animated bodies and living in accordance with the special qualities granted by the Creator, but that contradiction appears within both the animal and the image side of human nature and, when that occurs, the distortion from one side also distorts the other side. In actual existence, then, tension must be maintained within each aspect of the nature, both the animal and the image.

Humankind considered from the animal perspective means that a human being is a body, and is inconceivable as not body. The body is alive or animated, but there is no thought that what animates could be separated from body and regarded as intangible essence, with body as dispensable. In principle, then, the meeting of any genuine bodily need is legitimate, and is not in itself a manifestation of sin. By the same token, the desire to meet legitimate bodily needs is to be channeled rather than suppressed. But the fact is that in the conditions of existence the distinction between legitimate need and inordinate desire is always difficult to make both on an individual and a social scale, so that legitimate eating may become inordinate gluttony, and appropriate sexual expression may become something else. Even what we now call aggression, in the sense of movement through difficulties toward legitimate objectives, is acceptable, and gets off course only when it becomes inordinate. This whole position is well summed up in Paul's list of fifteen "sins of the flesh," in which the word used for "flesh" is the Greek *sarx* not the general word *soma* for body, and with virtually all of Paul's items representing (like gluttony) what we would now call attitudes or readinesses to act beyond proper limits.[4]

The image of God is a slippery metaphor, for it could wrongly suggest either a kind of die-stamp replica with the person as a little god or a robot reflection, or a mirroring of those characteristics of God that may not be consonant with human animal nature. Interpreters generally agree, although with degrees of emphasis on one quality or the other, that the real meaning of the metaphor lies in the relative capacity of human beings for freedom and for love. Freedom is understood more in the sense of self-transcendence than of self-direction, and therefore, although it is a good, it is an ambiguous good because its exercise may produce more harm than good in actual existence, and the more such power of freedom, the greater the potential harm as well as good.[5] Love is understood as the capacity for communication with God through worship and prayer, and also on the horizontal level with fellow human beings.[6] Clearly, the capacities for freedom and love in the conditions of existence are finite and limited. They may be misused both when their finitude is unadmitted and when their power is used to exploit either others or the self.

Whether the distortions come from the animated body aspect of humankind, or from that of the image of God, it is clear that the other aspect is automatically subverted at the same time. We can go even further and say that, even though distortions of the animal aspect may be serious also affecting the image aspect, the worst results occur from misuses of freedom and relationship both in themselves and in their effect upon the animated body aspect.[7] In other words, human beings go off course most disastrously through the way they exercise, so to speak, their highest capabilities. Although there may be frequent battles between lower and higher aspects of human nature, with victories sometimes one way and sometimes the other, those are basically skirmishes among guerilla troops. The decisive battles are among conflicting forces in the higher nature, which may draw the alleged lower nature into their respective strategic plans.

One other concept needs to be introduced at this point, especially as partially developed by Paul in the New Testament, namely, spirit as applied to human beings rather than to God.[8] Drawing both upon his Jewish heritage and the new life he believed he was experiencing through his understanding of Jesus Christ, Paul saw spirit (with a small "s") as the real unity and integrity of a human being, and as the gift of God rather than a human possession. As such, spirit related itself clearly to the image of God (freedom and love); but it was conceivable only with persons as animated bodies (including both *psyche* and *soma*). Thus, when Paul was attacked from both sides (by those believing in an intangible essence and those who thought the present body would be reconstructed) on his notion of the future of a human being after death, his answer was in the form of a "spiritual body," itself a paradox but not a contradiction. The body emphasis was attacking those who believed in an intangible essence that could discard body, and the spirit addressed to those who thought either that a body must be exactly as we know it or it is not a body, or to those who believed we possess our unity and hence know more details and time schedules than God has actually revealed to us.[9]

Paul's concept of spirit that includes body, but is not a human possession although the essential nature of human unity, was a bold step to show the proper relationship between humankind as animated bodies and as made in the image of God. Properly enough, even though it clarified the paradox, it remained paradoxical. Even though the Pauline approach has never been wholly lost in Christian history, Pelikan has shown how early came views that rejected the tension at least in part; and studies of later periods show how often the tension of the paradox was not sustained, with one result being that in common usage, even within the churches, spirit usually means something intangible as against the tangibility of body.[10]

HUMANITY AND DIVINITY IN JESUS CHRIST

From some time in the second century until the fifth, the principal focus of theological discussion was the relationship between the human and the divine in Jesus Christ. The answer that finally emerged at Chalcedon was a paradox. The notion of merging the two natures was rejected, and so were other tension-breaking solutions like keeping the two natures wholly separate. It was declared that Jesus had been fully human when he lived on earth at a particular time and place; but that he had been at the same time Son of God or divine in some way—and yet that any attempt to deny his full humanity (such as the reality of his death) by emphasizing his divine nature (as through his resurrection) was incorrect. The paradoxical nature of the position reached then is often misunderstood today because of the use of the term "person" to respect God in unity but in three "persons." The term must have come from the Greek word for the mask that actors were (*prosopon*, Greek; *persona*, Latin), even though that derivation did not cover the full intent of the early church leaders. It is important to remember that, in Greek theater, the wearing of what we call a mask was for purposes of recognition rather than concealment. Today some of us believe that the use of process rather than substantive modes of thought can do more justice to the values that the early theologians were trying to protect. Whether that is true or not, it was clearly their intent—especially for the purpose of the present discussion—not to permit the human nature of Jesus to be swallowed up by their belief in his divinity.

During those early centuries, the spread of Christianity was to the peoples whom the Jews called Gentiles, with only small groups among Jews. To the peoples of the Hellenistic world Christianity offered no problem because of the allegation of the divinity of Jesus Christ. It was his genuine humanity about which they had to be convinced. In sharp contrast, the Jews acknowledged Jesus as human, indeed as a fellow Jew. But they were offended by allegations of his divinity; and, with the exception of the few who became Christian, did not regard what Jesus had done as living up to their expectations about the Messiah who had been promised them in God's own time. Since Christians were increasingly Gentile, what early theologians fought for (after the first century) was the humanity of Jesus. That humanity tended to be lost in the Middle Ages and began to be recovered only during the Enlightenment of the eighteenth century.

It is obvious that Christians today, as over the ages, have looked upon the humanity of Jesus in a variety of ways, if indeed they believe in it genuinely or are instead "docetists" who regard him as appearing to be human while actually drawing divine aces out of his robe when he had a mind to. A catalogue of the varieties is neither possible nor necessary for purposes of the present discussion. What is noteworthy is that leading recent theologians as diverse as Karl Barth, Dietrich Bonhoeffer, and Paul Tillich have all looked to the human aspect of Jesus Christ as offering the best understanding we have of what genuine humanness is.[11] These three, and others, differ in their mode of approach. All, happily, rejected any forms of simple imitation. None believed that tensions and paradoxes would all be resolved if only we let Jesus be our guide in however sophisticated a fashion. But in their respective ways, each believed that our best though fallible understanding of the humanity of Jesus could be our guide or lodestar while we live within the paradoxes of actual existence. They all say in various ways that we do not learn what the truly human is and then decide that Jesus belongs to the species. Instead, we learn what Jesus was, and that becomes our criterion, in non-legalistic fashion, of what the genuinely human is. Such a view can of course be used to break the tension, as the so-called Jesus freaks do in an unsophisticated way, and others do in a more knowledgeable but unhelpful way. But so long as the traffic of reflection is between our understanding of humanness and our understanding of Jesus' humanity, without breaking the tension, Christian theology does offer a significant perspective on humanness even for those who are not Christians.

PILGRIM AND SAINT

Until our own century, the second most widely read book in Western Christendom, next only to the Bible, was John Bunyan's *The Pilgrim's Progress*.[12] Bunyan himself was a "tinker," or handyman, without formal education, who became a lay preacher, spent considerable time in prison for his oddities, and whose hallucinatory experiences at certain periods would have landed him in a mental hospital if today's standards had been applied. The careful analysis of Bunyan by the late Anton T. Boisen attempted to show, successfully in my judgment, that the dynamics that produced and resolved the mental illness were identical with those that provided the remarkably fresh religious insight.[13]

For our purposes only the basic skeleton of Bunyan's highly mythological story is required. The Pilgrim or Christian sets out upon the journey of life. He begins from the point where he has been shown the need for such a voyage. On the first leg he encounters difficulties and dangers mythologically portrayed. Sorely tried, he manages to get through them and to reach what turns out to be only a temporary resting spot. The brief time spent there is much like our modern understanding of the period of equilibrium between stages of human development. Then the Pilgrim moves on. New dangers are encountered. The Pilgrim is never sure he is going to make it through, but somehow he does, on to another temporary resting place, and then in due course to another leg of the journey. During the entire experience the Pilgrim has a certain limited vision of the goal, but it is never wholly clear. And more important, such vision of the goal as he has, even though it serves as a powerful motivation, is not taken in what psychoanalysts today would call an "over-determined" sense, that is, the Pilgrim is neither under the illusion that the intervening steps should be easy because of the goal's goodness nor reluctant to take the next dangerous steps because of the goal's beckoning.

John Bunyan wrote nearly a century before the term "development," so far as I can discover, was used even in biology in the restricted sense of the opposite of envelopment, or as a presumed automatic unfolding.[14] It was to require another century and a half before development was used in the modern sense, as a complex combination of biological, psychological, and social movements of the human being through various stages of life, including both the unfolding or push factors and those of cultural conditioning and group and personal decision. Modern developmental theory is very far from showing a uniform diagonal line on a graph.[15] It has ups and plateaus and downs and various combinations of these elements; and we are only on the threshold of serious study of such patterns in adult years.[16] Mythological as Bunyan's account was, his general outline seems both a foretoken of modern developmentalism and a challenge to examine more seriously the factors involved in the adult legs of the journey.

The Pilgrim's "progress," which term is not to be confused with the modern connotation of automatically becoming better, has some points of similarity but more of difference with the culturally universal myths of the hero, as conveniently summarized in Joseph Campbell's volume.[17] In most hero myths the central character leaves home at a particular time, roughly speaking when he becomes adult. Both the hero and the Pilgrim presumably have some choice about whether to move or stay, but if they did not move there would be no story.

The hero begins his journey, and ordinarily encounters dangers represented mythologically. Before each danger is confronted, he is unsure of victory; but after each victory, he gains a renewed and increased sense of strength which, fortunately for him, never leads to carelessness when the next dragon pops up. Usually a beautiful young woman turns up near the end of the journey, like the sleeping beauty aroused by the hero's kiss. The end of the travels turns out to be the hero's home; but his return to it is in a wholly different status from what he had when he left. He is now a king or prince or some one of high status, complete with spouse. At that point the hero myths stop. If the high status couple later need marriage counseling, or their oldest son refuses to leave home to make his heroic trip, or inflation threatens the economic base of the heroic family, or the heroine grows bored by too much domesticity, we never hear of it from these stories.

There is of course great insight in the hero myths. Undertaking the journey at all does require some decision as well as push. It is human to feel a bit stronger after a dragon has been slain or a drowning escaped. The building up of demonstrated confidence, on the basis of competence, and without destructive over-confidence, is legitimate at least within certain limits. The acquisition of a bride is also a positive point, and indeed the Bunyan story may most easily be faulted in terms of the Pilgrim's absence of a companion. Even the return home, but in an entirely different status, is a depiction of real human life, and this need not be understood only in terms of our achievement-oriented cultures.

The Pilgrim is, however, except for his dispensing with companionship and sexuality, a more profound depictor of the human situation than is the hero. He meets with a greater variety of dangers, and of different types, than does the hero. The confidence gained from meeting one kind of threat may have no relevance as quite a different kind of danger is encountered on the next leg. For the Pilgrim, all resting places are obviously temporary, even though temptations to remain are not absent. For the hero, the push to the next series of trials is more automatic. The Pilgrim does finally reach his goal, (the Celestial City), which had been only partially understood before; and that goal is the Celestial City but preceded by death, with death being interpreted as not inappropriate. Perhaps our own century's neglect of *The Pilgrim's Progress* is due not only to the extravagance of the metaphor but also to the rejection of death as appropriate within the meaning of the whole movement of life itself. Nevertheless, with all due respect to the hero, his story is based on a kind of "power of positive thinking" theology, while that of the Pilgrim is more deeply paradoxical. I would argue also that the Pilgrim's story is a more accurate prefiguring of our modern understanding of the complexities of human development including the adult and older years, especially when humanistic values are included, whether they be theologically based or otherwise.

One word of caution is needed in connection with pilgrimage. Over against John Bunyan there are the horrors of the Crusades of the Middle Ages, which represent not the genuine Pilgrim but a mixture of racial and religious prejudice, the most materialistic of adventurous efforts, and much else that is ethically indefensible. The pilgrimages to Canterbury described by Chaucer were neither heroic nor venturesome; and they were apparently dull except for the stories told at the nightly resting places. And with no discredit to Moslems, our most prominent literal pilgrims of today, the clarity of their objective and the ritualization of their procedures make their experience wholly different from that of Bunyan's Pilgrim, whatever merit their journeys may have on their own terms. The journey of Bunyan's Pilgrim never breaks the paradox, and therein lies its uniqueness.

Properly understood, the conception of the saint is as paradoxical as that of the pilgrim. Sanctity refers to holiness, or appropriateness from the perspective of what is felt to be divine, and not necessarily to goodness, especially in the moral sense. No doubt the idea comes from the primitive "mana" of many cultures, and which even in more sophisticated forms, is experienced as both luring and repelling, as Rudolf Otto summarized so clearly nearly sixty years ago.[18] When the ambivalence of holiness was translated as an apparently existing attribute to particular human beings, the resulting conferred power could be, and was, used both for purposes of healing and destruction. As sophistication and morality advanced, more cultural pressure was put on holy persons to exercise their powers for good only, and therefore the alleged saint was expected to use his holy condition for good, not evil.

The next stage was to reconsider the notion of whether any human beings could, regardless of their goodness, be considered more holy than any others. Instead, was not the total life of every person, if properly conducted, a movement in the direction of holiness as appropriate to him or her, with no one ever being fully holy in actual existence? Throughout its first three centuries Protestant Christianity wrestled with this question under the title not of "saint" but of "santification." Thus, long before process or developmental modes of thought were introduced into ordinary discourse, the gist of them was the foundation of discussion. The guidance of the course of one's whole life after one had presumably been saved was a concern exercised in many kinds of ways. Some communities tried to withdraw from the rest of the world. Others had such suspicion of holiness as an attribute of particular people that they insisted on no priests or ministers. Protestantism generally was negative toward monasticism. Despite many indefensible ways of trying to retain the idea of saint in terms of process and development, much of Protestantism did, until recently, retain the paradox (movement toward goal but also backward—no perfection) inherent in the conception. It should be noted also that the Roman Catholic tradition of "spiritual direction" was, in effect, a move in the same direction, however much it too was often subverted by various forms of legalism.

Today sainthood is out of fashion officially, except in those rare instances when the pope elevates some one dead long enough to elude inspection by the CIA or the KGB, or perhaps among Mormons in their secret councils. But the presumably old fashion of ascribing sainthood to special people is far from dead, and it is not confined to figures like Father Divine, nor even to persons who represent themselves as religious leaders. A very few persons, like Gandhi, survive almost universal inspection so that the appellation of "saint" is not considered out of place. But the fact is that, in the churches and out, the idea of the saint as moving toward a holiness never to be realized in actual existence, but regarding the process as in the interest of the self and others because of the benevolent nature of the divine will, has almost disappeared. Except in fundamentalism and other reactionary movements, where the prevailing attitude is that sainthood, under other names, has already been achieved or conferred, the only serious wrestling with the santification process seems to be among those concerned with an appropriately modern form of what is usually called "spirituality," as represented, for instance, in the writings of Thomas Merton and Henri J. M. Nouwen.[19]

The two ideas of pilgrimage and sanctification have been discussed primarily to show that, long before modern views of development and process had evolved, the Western theological tradition had insisted that any conception of desirable humanness, to use our current vocabulary, would be meaningless if the quality were seen only cross-sectionally, as simply present or absent, instead of as a paradox involving the whole living of a life. Translated into modern terms, and avoiding the confinement of these ideas to an explicitly theological frame of reference, the implication is that any general view of humanness that is not articulate about stages in the process of development may become an irrelevant abstraction. In theological history the transitional figure between Bunyan and the present is probably Kierkegaard, who turned the dragon into anxiety and sanctification into "becoming a Christian" but never arriving at that goal as an existential condition.[20] His being chosen as the father of modern existentialism (as if he were not first the great demonstrator of paradox) may have contributed to the still prevailing neglect of these basic notions of what desirable humanness, in the course of actual living, really is.

THE KINGDOM OF GOD: NOW AND THEN

As the discussion in the previous section tends to focus around the individual person, even though Everyman is implied, care must be taken lest the theological approach to desirable humanness be mistakenly thought to be unconcerned about society and the whole world. In Christian thought the central symbol of this concern has been the Kingdom of God or the Kingdom of Heaven, itself a paradoxical idea that seems to have been as close to the core of the teaching of Jesus as modern biblical scholarship can discover.[21]

Jesus seems to have believed and taught, on the one hand, that the kingdom is within, is present, has a new dimension. In modern language, such a declaration calls for assuming a certain attitude—not denying problems, evil, sin, or even obligation—but viewing them, and the more felicitous aspects of existence, from a new perspective. So far as we understand Jesus's ground for regarding the kingdom as actually present in some way and degree, it would appear to rest upon the deep conviction of the unambiguous love and benevolence of God toward human beings. When such a conviction is not a "power of positive thinking" cover-up of the negative realities of actual existence, then it is both produced by, and in turn produces, such qualities as faith, hope, and love. It is not the whole kingdom. Some interpreters have called it foretaste, which emphasizes the incompleteness correctly but wrongly minimizes the reality of the current power. To Jesus it seems to have been a current reality for some, a potential reality for all, the power of its contemporaneity in no way diminished by its not being everything that God plans or hopes for the future.

Jesus also taught that the complete kingdom is to come, in a time and a way known only to God. He is not reported as speculating about the time or the way. He is quoted as saying that the way may be contrary to our expectations, and that we should live as if the time might be any time. Like the Hebrew prophets, Jesus believed that we could safely leave time and way in the hands of the loving God, while our primary task is to accept the real though incomplete kingdom as it now exists. As to what he said or meant about altering the social order, scholars have expressed different opinions, even though there is agreement that he was neither an advocate of general social revolution nor of blessing the social *status quo*. What seems most evident is that the relation between the kingdom now and the kingdom then remained in proper paradoxical form. Accepting the kingdom now came first, and living accordingly. But the attitude enabling that acceptance could be both tried and supported by belief in the complete kingdom. Humans are to live as if it might come any hour and in the most unexpected way. But about that hour and way there is to be agnosticism, a confession not only that we do not know but that we do not need to know because God is what he is. It could be said, therefore, that the relation between the kingdom now and the kingdom then in Jesus is indeed paradoxical, but that the paradox contains both tension and mutual support.

In Western history almost every conceivable kind of relationship between religious groups and contemporary cultures has appeared, as carefully analyzed by H. Richard Niebuhr in *Christ and Culture*.[22] In regard to the United States, Niebuhr's *The Kingdom of God in America* is still unsurpassed in demonstrating the subtle variety of ways in which U.S. culture has been tempted, and has often fallen, by the notion that we are well along the road to the final kingdom even if we have not wholly arrived.[23] Some more recent studies of "civil religion" tend in the same direction although not all retain the complexities of Niebuhr's discussion.

Without suggesting that the rapidity of social and technological change in our time is not new and alarming, for it is, I am impressed by the fact that some commentators describe the "human being" in individual terms, and portray the obstacles to be overcome in collective and external terms. This apposition is remarkably similar to the New Testament's dealing with the Kingdom of God in the present tense as against its identification of the enemies as being the principalities and powers. So far as both diagnoses go, they seem in accord. But even as Jesus suggested that the forces to be combatted are internal as well as external, it is also the case that a technocratic mind-set and depersonalization of attitude within are just as much enemies as external technology and impersonal social institutions.

Is there, at least by implication, a similar kinship between the desirable (i.e., Kingdom of God) human condition in the present and in the future then? I would hope so. A conviction that true humanity, in the best though fallible sense in which we know it, is still present and contemporaneously active, may be our equivalent of the actual kingdom now. With that, it may be possible for us also to have some vision, properly unclear about time and specifics, about a kingdom then, which can be in proper tension with the present, and with the two convictions also providing mutual support. And precisely as the view of Jesus did not and could not separate the individual from the social dimensions, so I believe we must reason today.

HERESIES ABOUT HUMANNESS

In principle, all the preceding sections have argued that a heresy about humanness appears at any time any of the basic paradoxes, with their necessary tensions, is broken in either direction. If the term "heresy" smacks too much of the Inquisition, or of witch hunts ancient or modern, we can easily substitute terms like subversion, threat, distortion, falsification, or betrayal. Under whatever term, the implication is not of one idea alongside a different idea, but an aggressive twisting of some truth.

Many kinds of heresy may result when the tension is broken between the obviously existing human situation and the equally true conviction about the isness of created human nature as righteous and good. Heresy occurs when the dangers are denied or minimized, or when they are judged so overwhelming that even hope cannot be evoked. It appears also, however, when only human ideals and projections are believed to stand against the evils and dangers of the existing situation, and no credence is given to a created nature that has an isness different from the ambiguities of the observed beings at the surface level.

Similarly, heresy appears when the understanding of humans as animals is either abhorred or worshipped, when body and desire are either to be suppressed or given impulsive free rein. So with humans as made in the image of God: either when the alleged higher qualities are falsely taken as unambiguously good; or when they are regarded as too angelic to deserve human attention. Further, even when the proper tension is maintained between the idea of the human being as animated body and that of the human being as image of God, heresy may appear in breaking the paradox by which those two are related. The most common form of subversion is setting the alleged higher against lower natures, and heresy appears whichever option is taken. More subtle heresies appear in such forms as extreme other-worldliness or extreme this-worldliness, or indeed in more or less unqualified optimism or pessimism. Both optimism and pessimism are like prejudice, in that the metaphorical spectacles they wear are so colored that discriminating perception is made impossible.

Whether or not Jesus Christ, or some other figure or set of qualities, is regarded as our best picture of the true or desirable human being, there is heresy either when the ambiguities (along with the deficiencies in our understanding) are denied, or when true humanity is denied any transcendent quality so that approaching it seems to become a simple possibility. To put the matter in another way, genuine humanity cannot be humanly unrealizable without heresy (hence ambiguities will continue); but neither can it be worth the name if just a little more of something or other will produce all it has to offer. I have intentionally posed these heresies without explicit reference to God; for I believe the same dynamics apply to the unbeliever and the believer alike.

As to the pilgrim and the saint, the most obvious heresies come from arresting the journey, whether in the valley or on the mountain, in seeing only the threats and not also the oases or vice versa, in exaggerating our understanding of the goal or or falsely believing we have reached it, or in cryptic despair losing the will to move toward the goal at all. In our day it may be the shoppers after mental health, as if it were the highest good, who are inwardly afraid to become pilgrims moving toward a holiness that can never be achieved. It may be those trying to become "fully functioning persons" or "self-actualizing persons" who believe falsely that such five-fold or eight-fold paths guarantee them against the slings and arrows, and assure them that they have at least almost made it.

As to the kingdom, no other area today is producing so many heresies; and yet perhaps some of the heresies may be turned back into the non-heretical realm with bits of insight here and there. Futurologies come from three main groups: official commentators, who assure us the future is bright at least within a few months or years; the people, including us all at some times, who are mainly concerned with the future as well as the present price of potatoes or whatever else seems of immediate concern; and the "observers," religious and secular, who play it safe by pessimistic predictions so that they can be gratified at their wisdom if they are right and clear-thinking realists if they are wrong. My suspicion is that all three kind of futurologists are heavily tainted with heresies.

It seems no accident that the profusion of futurologies comes concomitantly with a rising popularity of what is called, among other names, the Human Potential Movement. Aspects of that trend are certainly not heretical. But far too many manifestations are based, as Binstock has noted, on the conviction that "emotionality is evidence of depth and genuineness of feeling."[24] That, Binstock continues, is an "illusion." He adds, " . . . they constitute collectively a special American therapeutic passion of the present day. They are described as 'experiential,' 'mind-expanding,' 'potential-enhancing,' 'primal,' 'genuine,' 'existential,' and 'real,' and, as an added attraction, they 'get the anger out.' "[25]

Neither futurology (or eschatology, as theology terms it) nor "contemporology" of feeling is inherently heretical. But each of these trends appears, in at least many forms today, to be reacting defensively against confronting the whole reality, over and above the distorted or exaggerated contents that the discussion has illustrated. Even though I cannot exclude many forms of these trends from the heretical category, it does seem that some of them could be bridges rather than dead-ends. Although not an avid reader of science fiction, I have found nevertheless in some of the best of it an imaginative use of the future that interacts with the present both as tension and mutual support. And in some aspects of modern therapeutic work I see reaching out beyond immediacies to, for example, a new depth in understanding hope in the face of genuine and warranted despair. A heresy need not remain such forever.

THEOLOGICAL CONSTRUCTION ABOUT HUMANNESS

This discussion has moved perspectively toward a central thesis, that Christian theological insights into the true nature of humanness can be understood only paradoxically. A paradox is an apparent contradiction from some points of view. To declare a paradox, however, requires another point of view that accepts the inherent tension and declares that it is not the last word, even when the tension in the present situation cannot be resolved. The perspectives used here to establish this thesis as characteristic of the best Christian thought have been man as animal and also man as made in the image of God, humanity and divinity in Jesus Christ, the pilgrim and the saint (each involving a paradox), and the kingdom of God as both a present and future condition. Each perspective discussed has been held to support and strengthen the thesis. Since it is in the nature of perspectives not to be exhaustive, those chosen have been only samples; but they are very important samples. It is implied that the use of other significant perspectives from the theological tradition, if properly selected, would further strengthen the thesis.

I have not held that theology has always been true to its paradoxical vision. In every age and among every group the tension of paradox has proved too difficult to sustain, and undue emphasis has been placed on one side of the paradox at the expense of the other whether in things great or small. Much of theological history is an account of these one-sidednesses, followed sooner or later by attempts at correction ordinarily themselves overdoing the correction at the beginning. There is of course no guarantee that the corrections will restore the full flavor of the original paradox. Furthermore, the social conditions always are different from those obtaining when the original insight appeared; so that the issue of what is fresh discovery, as against what is rediscovery of ancient intuition, is never easily solved. Pitfalls abound, and it is my position that there is no final externalized authority to substitute for the patient but fallible work of testing both by adherence to original intuitions and by viewing the fruits in the actual situation. Disappointing though such a position on theological authority may be to those who want one or another kind of certainty, it seems necessary if the paradox is to be maintained and not openly or cryptically denied.

It has already been noted that a genuine paradox cannot be declared unless there is a certain attitude in the subject toward the objects that appear to be in at least partial contradiction. That is, to assert paradox is to declare a kind of relationship between the human subject, and his perceptions of the human situation in what he regards as its proper context, that acknowledges the limitations upon his knowledge, upon his power, and even upon his love. A statement of paradox is, therefore, not simply about conditions out there. It is about a stance toward what is believed to be out there. And what is out there becomes acceptable in here—which of course need not imply merely conforming to any prevailing *status quo*.

The question is sometimes asked whether acceptance of the tension of paradox may lead to indifference, to passivity in the social or intellectual or even the feeling dimensions of life. It cannot be denied that, as a chronological consequence of confronting apparent contradictions, some persons or groups have fallen into a passive stance in one dimension of life or another. This is one way of looking at despair in the general sense in which Kierkegaard used the term. In such instances, it is usually clear that the tension of the paradox has been too great for person or group to bear. But a response of passivity, resignation or despair is in itself a breaking of the paradox; and if the paradox lies not merely out there but also in here, then it can be said that the passive respondents have not genuinely accepted paradox. They have only appeared to do so.

To take the situation as genuinely paradoxical is, on the one hand, to be prepared to live with its tension, but it is, on the other hand, to strive for untangling at any point where that is possible. There is active seeking without, however, the illusion that the search will ever break the tension. Thus, the acceptance of the big paradox may lead to resolution of many smaller paradoxes.

If space permitted, it would be interesting to examine various current theological trends in the light of the thesis. Some trends would be found trying to correct previous one-sided concerns and perhaps overshooting the mark. Others, weary of tension, would be seen as stressing one side at the expense of the other. Still others would be found supporting a superficial view of the paradox. Happily, yet others could be shown as gaining fresh insight into the profundity of paradox and relating it to the existing human and social situation.

Thus, the renewed interest in Jesus could be seen as an attempted corrective to the one-sided emphasis on Jesus as Christ. The resurgence of various forms of theological conservatism, usually combining privatism with authoritarianism, may be viewed as a retreat from the paradoxical tension. The so-called charismatic movement appears as an attempted correction against the hand of a tradition appearing cold or unemotional. The renewed interest in transcendence, however understood, attempts to correct against a unidimensional view of the world that did indeed deal one-sidedly with the paradox. Time does not permit elaboration of such trends, and their analysis in light of the thesis.

Specifically in relation to humanness, it is important to realize that theology would be breaking the paradox if it saw humanness alone and in itself, or in one-dimensional terms. At the beginning of his major work, John Calvin said that theology could begin either with God or with man, but that it remained theology only if each were seen at all times in light of the other. In modern language, that approach is a matter of context. It implies that what we know or believe about man is faulty unless seen in the context of God. It implies also that, since we are human beings, we know God only in relation to ourselves and not God in himself (which is a basic part of the meaning of revelation in theological discourse).

In both Catholic and Protestant circles, serious considerations of "Christian anthropology," the examination of humanness and the human situation, is much on the increase. The forerunner of this trend was the martyred German theologian, Dietrich Bonhoeffer.[26] He wrote of "man come of age," a human situation in which even the most sophisticated equivalents of praying for rain would no longer do, and human beings must assume responsibility for their situation. But he also referred to Jesus Christ as the "man for others," and "true man," thus indicating that it is not sufficient to understand our humanity by projecting our ideals, but we must instead find our pattern at another level of real historical existence.

Another major contributor to theological anthropology has been the French Catholic theologian and philosopher, Gabriel Marcel, especially in his understanding of the place and function of hope in genuine human beings.[27] In schematic form Marcel held: where there is expectation, hope is not needed; hope can arise only when there is a real *problematic* felt as some degree of despair; the experience of despair does not in itself guarantee the appearance of hope, even though the awareness of the despair may suggest that hoping has already begun; and the factors that produce hope, when it is needed, may vary, but they must in any event be regarded as fundamental to the entire outlook, and not peripheral or accidental. Understood in Marcel's way, hope does seem essential to genuine humanness. It is another way of looking at the human attitude required to grasp a paradox properly as such.

To date no complete theological anthropology has appeared that maintains proper paradox but corrects for the centuries-old suspicion that putting man in the foreground runs the risk of dethroning God or making superficially existing man the measure of all things. Perhaps no such masterwork will ever appear. Perhaps the new and detailed knowledge of man is and will remain so vast that only segments of it, along with the ancient paradoxical intuitions, can be mastered by any one. Perhaps the new theological anthropology must be a group enterprise, done from various perspectives, in some contrast to the past. But all such efforts, partial though they may be, will themselves endanger the paradox if they speak only of the human. The topic of theology is God in relation to man and man in relation to God, neither one without the context of the other, neither one without the paradox of tension and mutual support.

NOTES

1. Mircea Eliade, *The Sacred and the Profane* (New York: Harcourt, Brace, 1959); *Myths, Dreams and Mysteries* (London: Harvill Press, 1960); *Images and Symbols* (New York: Sheed and Ward, 1961).

2. Søren Kierkegaard, *The Concept of Dread* (Princeton: Princeton University Press, 1944).

3. Seward Hiltner, *Theological Dynamics* (Nashville: Abingdon Press, 1972), see chapter 4.

4. See Galatians 5.

5. See *Theological Dynamics*, chapter 1, on freedom.

6. See Daniel Day Williams, *The Spirit and Forms of Love* (New York: Harper and Row, 1968).

7. This point is made with force by Reinhold Niebuhr, *The Nature and Destiny of Man* (New York: Charles Scribner's Sons, 1941-43), especially vol. 1.

8. Paul's most extended discussion is found in 1 Corinthians 15.

9. The most celebrated recent discussion, despite its brevity, is by Oscar Cullmann, *Immortality of the Soul or Resurrection of the Body* (New York: The Macmillan Company, 1964).

10. Jaroslav Pelikan, *The Shape of Death* (Nashville: Abingdon Press, 1961).

11. As illustrative works by these authors, see Karl Barth, *The Humanity of God* (Richmond: John Knox Press, 1960) and his masterwork, *Church Dogmatics*, 13 vols. (Edinburgh: T & T Clark, 1936-69); Dietrich Bonhoeffer, *Ethics* (New York: The Macmillan Company, 1955), *Letters and Papers from Prison* (New York: The Macmillan Company, 1953), *Creation and Fall* (New York: The Macmillan Company, 1959), *The Way to Freedom* (New York: Harper and Row, 1966); Paul Tillich, *The New Being* (New York: Charles Scribner's Sons, 1955), *Systematic Theology*, 3 vols. (Chicago: The University of Chicago Press, 1951-63).

12. John Bunyan, *The Pilgrim's Progress* (New York: The Macmillan Company, 1948). This seems to be the most recent printing.

13. Anton T. Boisen, *The Exploration of the Inner World* (Chicago: Willett Clark and company, 1936, reprinted by Harper and Row).

14. Seward Hiltner, "Darwin and Religious Development," *The Journal of Religion* 40, no. 4 (October 1960):282-95.

15. See Erik H. Erikson, *Childhood and Society* (New York: W. W. Norton and Company, 1950).

16. See Don S. Browning, *Generative Man: Psychoanalytic Perspectives* (Philadelphia: The Westminster Press, 1973).

17. Joseph Campbell, *The Hero With a Thousand Faces* (New York: Pantheon Books, 1961).

18. Rudolf Otto, *The Idea of the Holy* (New York: Oxford University Press, 1923).

19. See Thomas Merton, *The Seven Storey Mountain* (New York: Doubleday and Company, 1948), *The New Man* (New York: New American Library, 1961), and *Contemplative Prayer* (New York: Herder and Herder, 1969); Henri J. M. Nouwen, *Creative Ministry* (New York: Doubleday and Company, 1971), *The Wounded Healer* (New York: Doubleday and Company, 1972), *Out of Solitude* (Notre Dame: Ave Maria Press, 1974), *Reaching Out* (New York: Doubleday and Company, 1975).

20. Søren Kierkegaard, *The Concept of Dread*. See also *Stages on Life's Way* (Princeton: Princeton University Press, 1940), *Training in Christianity* (Princeton: Princeton University Press, 1952), *Either/Or* (Princeton: Princeton University Press, 1944).

21. See, for instance, Amos Wilder, *Eschatology and Ethics in the Teachings of Jesus*, rev. ed. (New York: Harper and Row, 1950).

22. H. Richard Niebuhr, *Christ and Culture* (New York: Harper and Row, 1951).

23. H. Richard Niebuhr, *The Kingdom of God in America* (Chicago: Willett, Clark and Company, 1937).

24. William A. Binstock, "Purgation through Pity and Terror," *The International Journal of Psycho-Analysis* 54, no. 4 (1973):500.

25. *Ibid.*

26. Dietrich Bonhoeffer, see note 11.

27. Gabriel Marcel, *Homo Viator* (New York: Harper and Row, 1962).

CHAPTER FOUR

The Religious Dilemmas of a Scientific Culture: The Interface of Technology, History and Religion

Langdon Gilkey

Our title may well seem puzzling. We can certainly understand that a scientific culture poses dilemmas for traditional religion of any sort. This has been assumed ever since our culture became scientific in the sixteenth and seventeenth centuries, and it became a virtual certainty in the nineteenth. But can a scientific culture as it develops raise its own religious dilemmas and show itself to be in need of religion in the way agricultural and nomadic societies were? This is the question I would like to investigate. We shall begin by exploring a middle term, history, and our understanding of history. For science has greatly influenced our sense of history, of where we are all going—and wishes to do so. And with the question of the meaning of history, religion inevitably enters the scene.

THE SCIENTIFIC/TECHNOLOGICAL PHILOSOPHY OF HISTORY

Generally speaking, scientists and technologists have not been directly concerned with philosophical questions about history and its meaning. In fact the general effect of a scientific culture has been to regard speculative philosophy, and especially philosophy of history, as about as "mythical" and full of fancy as religion and theology—and both as quite unnecessary for intelligent understanding. Nevertheless, the scientific and technological communities, despite their best intentions, so to speak, have generated out of their own abilities, commitments and hopes, a new understanding of history both reflectively and existentially, both in our thoughts and in our feelings. Indeed, almost any book by a scientist, when it speaks of the importance of science or of its role in society, reflects a particular understanding of history common to the scientific community. A philosophy of history, therefore, has been and continues to be created by science not as a result of its direct inquiries, not as a scientific hypothesis experimentally tested and as such a part of the body of scientific knowledge. Rather it is created when the scientific community thinks of its own role in history, reflects on itself and its knowledge, and sees itself through that knowledge as a creative force in history. Such a philosophy of history—and a very hopeful one—has been a presupposition, an important spiritual foundation, of the modern scientific community since its beginnings in the sixteenth century. The shattering of that spiritual base in our day has been, therefore, a crisis for the scientific community itself and for the technological culture it helped to create.

As Francis Bacon, the father of this understanding of science and of history, reiterated: greater knowledge (empirical not speculative knowledge) leads to greater control. When men and women know the way things around them work, then they can make those things work for them. Since science leads to far greater understanding of the dynamic causes of things, science is the secret of human control over the world. Thus with the advent of the scientific method—the deliberate, organized and successful effort to *know*—a new day will dawn for mankind: a day of new power, the power to control and direct, and so the power to remake the world, the power at last to realize human purposes through intelligent inquiry and the technical control which intelligence brings. Bacon's simple empiricistic understanding of the method of science has long since been superseded; but his vision of the role of science in human society and so in history has remained as the fundamental belief and hope of the modern scientific and technological world. Through this method we now know how to know, and through that knowledge to control. And as Dewey was to point out, the two—knowing and controlling—are in the end one and the same thing, the same power of organized intelligence, as he put it, to remake its world. Science functions as the means to human power, power over nature, society, and men and women alike. Thus both old and developing nations have seen scientific and technological knowledge as the keys to their military power and to their economic and social well-being—and each capital becomes fearful for its security if its "lead" in pure science is threatened.

There are many evils from which we humans suffer, and almost as many interpretations of those ills: which ones are basic and which peripheral. Generally most profound religions have interpreted the fundamental ills as coming from the inside of the self and from its finitude: from desire, pride, disloyalty, lust, mortality and death. Not so modernity. For most moderns our ills have come not from inside ourselves but from external threats: from our subjection to external forces beyond our present control and so from our inability, in being ignorant of these forces and how they work, to control them. The disasters of nature: floods and draught; the problems of heat and cold; the vast spaces of nature to traverse; the difficulties of communication; the diseases of the body; the paucity of essential goods: clothing, houses, heating, lighting and so on—these have been for modern technological man the basic "problems" that beset us in life. Since these are the main evils we face, then, if through our knowledge we can develop tools or instruments with which to deal with them, if we can control these "fates" of disaster, hunger, disease and want that afflict us from the outside, will we not be happy? Not only then can more of us survive, but through the technologizing of industry, we can survive securely, full and well—and a new kind of human life will be possible. *Homo faber,* the tool maker, transformed through science into technological and industrial man, is the "authentic" man because he alone can eradicate evil and bring in a new authentic world. Of course I have left out here those two other great issues of modernity relevant to human well-being: the distribution and control of political power and the distribution of economic goods and property. My point is, however, that whatever political or economic system we choose, this trust in science, technology and industrialism, this confidence in verified knowledge, technical know-how and organizational techniques, to cure human ills has dominated the scene, whether we look at Europe and North America, at Russia or Rumania, at India, China or Japan. The ultimate, long-term faith of this modern culture which began in Europe was in the scientist and his knowledge; but its immediate and practical hope, one may say, lay in the engineer: the builder of roads and factories, cars, tractors and planes, apartments and cities, sanitation, detergents and improved fertilizers. Through technological and industrial expansion jobs are created, wants appeased, politics made stable, food production increased, economic life rendered solvent, and so the power and self-determination of a people guaranteed. We all know this to be in large part true, and, when we are honest, we admit that we too welcome its results. Human life on our planet can hardly survive, let alone be pleasant, without these three: science, technology and industrialism.

Modern science and technology, then, brought with them, both reflectively and in our feelings, a message of promise for the future. A new and better world was now possible. Thus in the seventeenth and eighteenth centuries the theory of progress arose—slowly among philosophers and intellectuals generally and as a cultural mood slowly shared by all—out of the role science and technology were beginning to play and promised to play in society. The "future" eventually appears on a wide scale as an important category of thought, as the "place" where this "new" will appear, as the place where the ills of the past and of the present will dissolve. Science and technology produced that understanding of history which has dominated our entire present—capitalist and marxist alike—and has spread like fire across the globe wherever this culture has gone: the sense of an open future, of a future that will be better.

Out of this confidence in progress has emerged the new historical consciousness characteristic of modern culture: a sense of man's freedom in history to remake his world, of the possibility of the conquest of fate and evil, of human potentiality to fulfill itself and its life without either divine hope or a relation to eternity. No wonder modernity has to a new generation excited with this new knowledge and new techniques seemed to make traditional religion irrelevant and unecessary in East and West alike. Science seemed to show that religion was incredible, a result of man's childish ignorance when he did not yet understand his world or his own powers in it. But perhaps more important, science made religious salvation irrelevant now that the weakness of man in achieving his desires had through knowledge been changed into power, the power to control whatever he wills to control.

Within the new historical consciousness a new sense of change appeared. Men and women have always been aware of changes in nature and in themselves: the cycle of the seasons around them, and the pattern of birth, growth, and death in all that lives. Also, they have often been aware of confusion, chaos, and the loss of all stability in their social world—as at the end of the Roman Empire or in the feudal disorder before the Tokugawa period in Japan. But they had not been explicitly aware of a changing social world, of a transformation of the forms of life leading to something new, until modern times. Such awareness arose partly through the cataclysmic political events that overturned a seemingly changeless order: like the French Revolution in Europe and the Meiji restoration in Japan. But the deep modern awareness of steady and cumulative historical change has arisen, I believe, through the accelerated changes that technology and industrialism have effected in all our lands. Each of us sees, and feels deeply, the inexorable disappearance of the old and the appearance of the new in our social environment. Almost as if we were peering out of a speeding locomotive, the world we grew up in and felt at home in flashes by at lightning speed to be replaced by new scenery, by a new world. And with that transformation of the environment by expanding cities, exploding factories and new roads have come equally drastic changes in modes of life, in social relations, in the roles we each play in our world. We know in a new way that we live in a historical process, a process of the steady change of the forms of our social environment and so of ourselves. We are conscious in a quite new way of being immersed in history, and thus of facing with tomorrow a world we may not expect or even want. It is no surprise that modern philosophy—of almost all sorts, naturalistic, idealistic, existential—has emphasized process, change and the temporality of being as opposed to the eternity and changelessness of being. This has been the modern experience of whatever being we have known, and our philosophies have expressed this sense of the historicity of all that is.

For most people—except for those given privileges by the old order—it was with relief, joy and expectation that the world of yesterday was disappearing and a new world of tomorrow was coming. Thus when this sense of change first appeared, it felt good: change was promise, the promise of a new that would be better. To be in process feels good if process equals growth and progress. Change has a different feel, however, if we are not too sure whether the new will be better or not; and it is terrifying if the new appears as menacing. In any case, it is evident once again how science, technology, and industrialism as the main agents of change in our social world have generated feelings and reflections about history. They have together helped to create our historical consciousness, our awareness of our historicity and temporality: that we are what we are in historical process, that we are immersed in social change, and that we can through intelligence and will refashion, shape, and direct that change. Despite all its positivism and empiricism, its impatience with speculative philosophy and theology, modernity at the deepest level has been founded on a new philosophy of history: a philosophy built on faith in knowledge and its power to control, on the triumph through knowledge of human purposes over blind fate, and so on the confidence that change—if guided by intelligence informed by inquiry—can realize human fulfillment in this life. Such a view of history as guided by science and shaped by technology was the implicit "religion" of the West until yesterday.

THE ANXIETY AND AMBIGUITY ATTENDING SCIENCE/TECHNOLOGY

A change both in mood and in reflection, in feelings and in explicit thought, has occurred in the last decades with regard to this fundamental confidence in science and technology, and all that they imply about freedom, history and the future—like a sudden cover of storm clouds shutting out the bright sun. A chill, thematized in art, drama, novels and films and felt, if not thematized by most people, has settled over much of the West. The scientific community in particular is uncertain in an unprecedented way about its role and worried about its future and the future of the society it helped to create. Such anxiety appears whenever a "religious" confidence becomes shaky. The center of this new *Angst* is, I believe, a new intuition of the *ambiguity* of science and technology as forces in history. This is not primarily an uncertainty about the validity of scientific knowledge or about the reliability of technological skills. About these there are few new doubts—except among small (but growing) mystical and religious communities in the counter-culture. It is rather a radical doubt about their "saving" character and an anxious feeling that they create as many new problems and dilemmas for human life as they resolve, and even that they compound our ills rather than dissolve them.

Beneath this anxiety, but rarely explicitly expressed, lie deeper and more devastating questions. If a valid science and a reliable technology can really compound our problems rather than dissolve them, what does *that* mean about man and about the history he helps to create? Do we really increase our dilemmas by using our intelligence, our inquiry, our techniques? What does *that* mean about us? When these questions are asked, it becomes evident that the *user* of knowledge and technology, and so man himself, is the cause of this ambiguity. Possibly knowledge, informed intelligence and the freedom to enact human purposes that they give are not enough. Something seems to be radically wrong with the ways we use our intelligence, our knowledge, and with the ways we enact our control. Can it be true that human creativity, in which we have so deeply believed, is in some strange way self-destructive, that there is in human freedom an element of the "demonic," and that intelligence and informed freedom, far from exorcising the fates of history, can create their own forms of fate over which they also have no control? As is evident, all the great philosophical and especially religious problems about human life are implicitly raised here, problems unanswerable by science and unresolvable by technology, and yet raised by both of them the moment the future they seem to create becomes apparently oppressive and menacing rather than bright and promising.

As we all know, these deeper questions about scientific knowledge and control have been brewing for some time. They began with the development and use of the terrible new weapons and the threat to human life itself which the technological power evidenced in those weapons represented. These questions continued with the realization that technology provides the political authorities and a potential scientific elite with new and dangerous powers over ordinary people: political powers based not only on weapons and communications systems unavailable to the people, but also on the possibility of psychological and even genetic control of entire populations. Technology seemed now not so much to guarantee freedom and self-determination, individuality of life-style and privacy of personal existence, freedom from *natural* fates and freedom for becoming human, but rather it seemed to open up the possibility of an all-encompassing totalitarianism that could crush individuality and humanity, a possibility in which the human would be subordinated to a new kind of *social* and *historical* fate. These fears have been expressed for some decades in the Western consciousness—for example by Huxley and Orwell. However, two new factors have recently become visible that have widely increased this uneasiness about a technological culture: one of them since World War II and the other in the last decade.

THE DEHUMANIZATION ATTENDING
A TECHNOLOGICAL CULTURE

The first can be referred to as the dehumanizing effects of a technological culture. As Jacques Ellul has pointed out, technology is not only a matter of tools, instruments, machines and computers. It also characterizes a society in so far as it is organized, systematized or rationalized into an efficient organization: as in an army, an efficient business or a bureaucracy. Here all the human parts are integrated with each other into a practical, efficient , smooth-running organization where no time, effort or materials are wasted, where the product or the service is quickly, correctly and inexpensively created, and where a minimum of loss, error and cross-purpose is achieved. Thus are homes put up all alike by a single company and according to a single plan—for efficiency's sake. Thus is local government submerged in national bureaucracy. Thus do individual farms give way to farming combines. Thus is every small industry swallowed up by large, unified business or state concerns. The beneficial results of this technologizing or rationalizing of society are obvious: the rising standards of living of America, Europe and Japan have directly depended on the development of this sort of efficient, centralized administration of industry, distribution, services and government. And every developing country seeks to increase as rapidly as possible its rationalization of production and organization in order to feed, clothe, house and defend its people.

In the midst of these benefits, however, there have appeared other, negative consequences. As every advanced technological society has discovered, human beings are now not so much masters as the servants of the organizations they have created, servants in the sense that they find themselves "caught" and rendered inwardly helpless within the system in so far as they participate in it at all. By this I mean that they experience their personness, their individuality, their unique gifts, creativity and joy, their sense of their own being and worth sacrificed to the common systematic effort—an effort in which all that their own thought and ingenuity can contribute is to devise more practical means to an uncriticized end. Any considerations they might raise concerning creativity, aesthetics or the moral meaning of what is being done, any suggestions that might compromise the efficiency, the smooth-running of the whole team, are ipso facto "impracticable" and so by these standards irrational. Thus does the individuality of each lose its transcendence over the system; individual minds and consciences cease to be masters and become servants, devoted only to the harmony and success of the system. Human beings are present and are creative, but only as parts of a system; their worth is judged only with regard to their contribution as an efficient part; they are lured into being merely *parts* of a machine.

Society as a unified system has, moreover, proved ruthlessly destructive of many of the other, less public grounds of our identity as persons. It uproots us from that in which much of our identity, or sense of it, is founded, namely, our identification with a particular place and with a particular community. For it gathers us into ever larger groups of people similarly organiz-ed, and then it moves us about from here to there, from these people to those, within the larger society. It rewards and satisfies us only externally by giving us things to consume or to watch. After all, such things are all that efficient organization can produce. Having dampened our creative activity in the world into the rote work expected of a mere part of a system, it now smothers the intensity of our private enjoyments by offering us the passive pleasures of mere consumption. Thus does it stifle our inwardness.

Ironically the West had in its spiritual career discovered and emphasized, as had no other culture, the reality, uniqueness and value of the inwardness of each human being, of what was once called the "soul." But a concurrent theme, its affirmation of the goodness of life, the intelligibility of the world and the possibility through knowledge of the latter's manipulation and control, has gradually achieved an almost exclusive dominance. The combination of these two themes had promised to reshape human existence in relation both to nature and to the forms of social life, culminating in technology, democracy and socialism. Thus in comparison with the Eastern world, the West had creatively learned to manipulate the external, objective world and done much to humanize and rationalize the objective social order. But it has in the process endangered its own inward soul, the reality and creativity of the spirit. Thus having through science, technology, democracy and socialism helped to rescue the Orient's social orders, it now must turn back to the Orient in order to rediscover its own inwardness. And it is doing so in great numbers—ironically just when the Orient is itself grasping after the lures of Western technology and external progress!

Technological society promised to free the individual from crushing work, from scarcity, disease and want, to free him to become himself by dispensing with these external fates. In many ways, on the contrary, it has emptied (or threatens to do so) rather than freed the self by placing each person in a homogeneous environment, setting him as a replaceable part within an organized system, and satisfying his external wants rather than energizing his creative powers. Thus appears the first paradox: the organization of modern society necessary to the survival and well-being of the race seems now to menace the humanity, the inwardness and the creativity, of the race. In seeking to live by means of a surplus of goods unknown before and for the sake of such goods, we have found that men and women are in danger of losing themselves inwardly and so of dying in the process. What had been seen clearly with regard to individual life by the wisdom of almost every religious tradition, has been proved objectively on a vast scale by modern consumer culture: men and women cannot live by bread alone.

THE ECOLOGICAL CRISIS ATTENDING ADVANCED TECHNOLOGY

Consciousness of the second menacing face of technology is astoundingly recent, within the last half-decade. This may be termed the "ecology" crisis in its widest connotations. It refers not only to the problems of technological and industrial pollution of the water, air and earth and the despoilation of whatever natural beauties are left—though these are serious enough problems, and with energy and resources short will only get worse! It refers centrally to the exhaustion through expanded industrial production of the earth's available resources, in the end a far more serious problem. Medicine and greater production of food have increased the population; technology in both agriculture and industry has at an accelerating pace increased our use of nature's resources of fuels, metals and chemicals. In order to feed and care for that mounting population, such agricultural and industrial growth must itself expand almost exponentially. And yet if it does, an absolute limit or term will soon be reached; these resources will come to an end, if not in two or three generations, then surely in four or five. The seemingly infinite expansion of civilization and its needs is in collision course with the obstinate finitude of available nature and threatens to engulf both civilization and nature. For the first time man's freedom in history menaces not only his fellow humans but nature as well. In the past, with the development of the techniques of civilization, history was freed from the overwhelming power of nature and its cycles and submitted nature to her own control. Now civilization and history have become so dominant in their power that they threaten to engulf nature in their own ambiguity.

In this case that ambiguity is very great. A world economy, whether its domestic forms be socialist or capitalist, facing the combination of expanded populations and both depleted and diminishing resources, is a world facing even more bitter rivalries and conflicts than the past has known. It is also, ironically, a world facing in new forms precisely those "fates" from which technology had promised to save us: scarcity, crowding, want and undue authority. If there are to be rational solutions to these problems impinging on us from the future, and there are, they will require an immense increase in corporate planning and control on a world scale: control of technological developments, of the industrial use of natural resources, of distribution, of the wide disparities in the use and consumption of resources. Freedom of experiment, freedom for new and radical thoughts and techniques, freedom for individual life-styles, may well be unaffordable luxuries in that age. Perhaps most important, such rational and peaceful solutions will require from the nations with power an extraordinary self-restraint in the use of their power, a willingness to sacrifice their affluence lest they be tempted to use their power to grab all that is left for the sake of that affluence. All of this bespeaks an increase of authority in our future undreamed of in the technological utopias of the recent past. Whether we desire it or not, we seem headed for a less free, less affluent, less individualistic, less dynamic and less innovative world. The long-term results of science and technology seem ironically to be bringing about anything but the individualistic, creative, secure world they originally promised. In fact this progressive, dynamic, innovative civilization seems to be in the process of generating its own antithesis: a stable, even stagnant society with an iron structure of rationality and authority, with a minimum of goods, of self-determination, of intellectual and personal freedom. Such a new and grim world is by no means a certainty, for nothing in history is fated. But unless our public life—technological, political, and economic—is directed by more reason and more self-sacrifice than in the past, such a future has a disturbingly high probability.

THE AMBIGUITIES OF TECHNOLOGY AND THE AMBIGUITIES OF FREEDOM

As the hopes latent in science and technology gave birth despite themselves to a new understanding of history, so the new sense of their ambiguity raises for us a host of unavoidable questions about history and the relation of human freedom—of human intelligence, will and creativity—to history. Along with language, technology is in itself one of the most vivid manifestations of human freedom over its immediate environment. And, as we have seen, its growth in modernity has sparked the consciousness of that freedom in history, the ability of man to remake his world. And yet paradoxically, technology seems not so much creative of the freedom it represents as destructive of it—for it seems to be creating conditions which will of necessity absorb freedom into authority. Here the exercise of technological freedom, in order to remove the fates that determine freedom from the outside, has *itself* become a fate that menaces freedom—a strange ending.

This most vivid manifestation of freedom has exacerbated and not resolved the ambiguity of freedom, or of our use of it. In fact, technology, the creature of our freedom, has revealed the continual presence of what we can only call the demonic in history, the way our freedom is itself estranged and strangely bound—an old religious concept. For what clearly is amiss here, what reintroduces the ills we had thought almost banished, is not our intelligence, inquiry and technology per se; our creativity in itself is not at fault. Rather it is the demonic use to which it is put. At fault, as our religious traditions have emphasized, are the infinite desire and concupiscence, the greed and selfishness which motivate our use of scientific intelligence and technological power, which drive the infinity of industrial expansion that in turn ravishes and desecrates nature, and spurs us all to rivalry, conflict and doom. Under and behind the creativity of man, recently so clear to the modern West as the principle of historical salvation, lies the estrangement and so the demonic principle within man—whatever may be his ideals, his loyalties, his courage and ingenuity. Finally, and most ironical of all, man as the tool maker, as inquirer and technologist, has by modern savants been regarded as the paradigm of survival. He, not religious, mystical or mythical man, was the "practical" one who alone could handle "reality." Strangely now *homo faber*, as technologist supreme, seems himself to be alienated from "reality," bringing about through his technology his own self-destruction and showing himself to be the primary danger to the survival of his race. No more startling contradiction to the spirit of modernity from the Enlightenment to the present could be conceived.

Thus anew has what we can only call the *mystery* of history and of temporal being revealed itself to us, along with the potentiality of meaninglessness in the human story, as well as in individual life. Human creativity—yes, even informed intelligence and good purposes—is not simply "god" bringing to us unadulterated blessings, the answers to our every wish. With our creativity freeing us from old fates comes fate in a new form; with our creativity the demonic seems to be continually reintroduced into history. We live in a far stranger and more disturbing history than we thought, where even our apparent victories, our most cherished mastery, our greatest intellectual and practical triumphs, help to seal our doom!

THE AMBIGUITIES OF TECHNOLOGY AND RELIGIOUS RESPONSE

I need not in conclusion underline that these paradoxes arising out of the role of science and technology in modern life raise religious issues. It is obvious that all these questions make direct contact with the themes, meanings, questions and answers of speculative philosophy and high religion. If it is the way we use our creativity, our intelligence and freedom—not our lack of them—that is at fault, then is there any recourse for us from this estrangement of our own most treasured and precious powers, from this bondage of our wills to self-destruction? We seem to need rescue not so much from our ignorance and our weakness as from our own creative strength—not so that either our creativity or intelligence is lost, but so that their self-destructive power is gone. Thus the religious question of a transcendent *ground* of renewal, not from ourselves but from beyond ourselves, is raised by the most impressive of modernity's achievements—its scientific intelligence and its technological capacities. The creative role of religion is not to replace intelligence and technology with something else, but to enable us to be more intelligent, more rational, more self-controlled, and more just in our use of them. Further, if it is our use of creativity which threatens the meaning of our history—because it renders ambiguous our common future—then again the question of a meaning in *history* which is more than meaning which we can create or give to history appears. In the face of the fate with which our own creativity seems able to dominate us, the religious question arises whether there is any other providence that can rule the fates that seem to rule over us. Our history and our future are not threatened by the stars or the blind gods—by forces beyond us. Ironically they are threatened by a fate which our own freedom and ingenuity have themselves created. Here too, therefore, for us to be able to face our future with confidence— for we can no more live without technology than we can apparently live humanely with it—we must trust in a power that tempers and transmutes the evil that is in our every good and the unreason that is in our highest intelligence.

Such issues as these, raised not against science and technology but precisely by them, cannot be understood or even discussed without religious categories. Moreover, on the existential as well as on the reflective level, they cannot be handled without a confidence and a trust born of religion. The anxieties involved in facing such a potentially menacing future require the serenity, the courage and the willingness to sacrifice which only touch with the transcendent can bring. Modern culture in the development of its science and technology has not made religion irrelevant. It has made religious understanding and the religious spirit more necessary than ever if we are to be human and even if we are to survive. Technology by itself, technical-manipulative reason, if made the exclusive form of reason and of creativity, has been clearly shown to possess a built-in element that leads to its own destruction and the destruction of all it manipulates. It must be complemented by the religious dimension of man and by the participating, uniting function of reason if it, and we, are to survive at all.

Specifically, science, technology and the society they constitute must be tempered and shaped by the religious dimension of man, not with regard to their own modes of inquiry, their conclusions or even their specific programs— though the latter do need ethical as well as "practical" assessment; rather this tempering and shaping has to do with the humans who use them and on whom they are used. From religion alone has traditionally come the concern with the human that can prevent the manipulation of men and the dehumanization of society; and from religion alone can come the vision or conception of the human that can creatively guide social policy. From religious confidence alone has come the courage in the face of fate and despair—especially when these two arise from the distortion of *our own* creativity—concerning a future that will by no means be easier than the past. For humanism can count on only our own deepest creativity; when that too reveals itself as ambiguous, then despair and cynicism rather than humanistic confidence appear. From religion alone can come the healing of desire and concupiscence, that demonic driving force behind our use of technology that ravishes the world. And from religion alone can come a new understanding of the unity of nature, history and mankind—not in human subservience to nature and her cycles, but in an attitude which, recognizing the unique spiritual creativity of mankind, can still find human life a dependent part of a larger spiritual whole that includes the natural world on which we depend. Such a unity with nature has been expressed in much traditional religion, especially in the Orient. It must be re-expressed and reintegrated in the light of the modern consciousness of human freedom, technology and of history. Naturalistic humanism cannot achieve such a unity with nature through spirit. Without the category of the ultimate, the transcendent, or the divine beyond and yet inclusive of both nature and human being, man is either subordinated to nature or, recognizing his transcendence, uses her for his own "superior" ends. Thus religion is necessary in a technological society if such a society and the nature on which it depends are to survive.

But—and religion both East and West should take note—it is only a religion related to history, to social existence and to the human in its social and historical context that can complement, shape and temper technology. A religion that lifts us out of time or gives us only individual peace, that vacates society and history in favor of transcendence alone, will only encourage an irresponsible and so a demonic technology and will foster and not conquer a sense of fate within history. We are, whether we will or not, *in* history, immersed in historical and social process; and here our lives for good or ill are led. On our response to the social and historical destiny of our time—in this case a technological destiny of vast ambiguity—rests the validity and meaning of our inward spiritual life, of our religion. Only a religion that responds to a transcendence beyond our own self-destructive powers and yet that finds its task centered in our common historical and social future can become a genuine means of grace to us.

CHAPTER FIVE

Two Truths: The View from the Social Sciences

Daniel Yankelovich

Being human in a society dominated by technology and impersonal institutions is a broad topic, and it can be interpreted in many ways. I interpret it as a question raised with a slightly skeptical overtone: Is it truly possible to be human in a society dominated by technology and impersonal social institutions? In responding to this question I will first look at some of the difficulties that social science has in wrestling with this kind of broad humanistic problem. Second, I will do the best I can, with the help of material from the social sciences, to engage the question. Finally, I will conclude with several broad generalizations.

No one, of course, can pretend to speak for all, or indeed for any one, of the social sciences. Clearly, therefore, this has to be a personal and idiosyncratic point of view. From that point of view, I believe that the social scientist finds it difficult to bring a wise and prudent judgment to bear on the broad theme of this topic because of a fundamental disarray at the heart of the social science enterprise. To answer the question, it would be necessary to bring into proper harmony two truths which are now held to be mutually exclusive by two warring camps within the social sciences.

TWO TRUTHS: THE NATURE/NURTURE CONTROVERSY

One truth is embodied in the environmentalist perspective represented by the dominant school of five "in" social sciences: behavioral psychology, economics, political science, sociology, and even anthropology. This perspective advocates the truth that to be human is to exist in society and in culture and to be shaped essentially by social interaction. In the eighteenth and nineteenth centuries it used to be popular to speculate on what someone such as a wolf-boy would be like who was brought up outside of society and culture—nurtured perhaps by animals—but in any event reared apart from any form of culture. There was much speculation, even some degree of experimentation. From the point of view of the environmentalist truth, the wolf-boy concept—the notion of someone who exists outside of culture and society—is simply a misleading abstraction. Such a creature would not be recognized as human. He might be a creature, but he would not be a human being. That is one perspective, one truth.

The second truth is embodied in the evolutionary human-nature perspective that holds that even though society and culture may be necessary to elicit our human possibilities, people are not infinitely plastic and malleable. We bring with us into the world an inherent and universal human nature representing seventy million years of evolution that makes it meaningful to be able to say about any society or any culture that such and such a society either is consonant with or, conversely, is constrictive or oppressive of fundamental human needs. This is decidedly a minority point of view in the social sciences held by some ethnologists, some psychoanalysts, biologists, and a minority of psychologists and anthropologists. The perspective advanced by this truth focuses on varying stages of human development, emotional as well as physical. Professor Hiltner stressed this point, emphasizing the push factor and the essentiality of a developmental concept in the identification of needs, motives, and characteristics with respect to different stages of development that are common to all humanity.

The difficulty, perhaps even the fatal flaw, in social science is that it has not been able to bring these two truths together, the assumption being that if you hold one you cannot hold the other. For several powerful reasons these two truths are either held to be incompatible or else are joined together in some bland and artificial manner such as the formula that heredity and environment interact without specifying how they interact, what the invariant developmental stages are, what the fundamental human needs and characteristics are, what kinds of society are good for people, and what kinds of society are bad for people. It would take too long to explain in detail the reasons for this impasse, although from the historical point of view, they are fascinating. Let me simply mention three of the critical resistances which combine to support the majority environmentalist position and work against integrating the human-nature perspective.

CRITICAL RESISTANCES TO THE INTEGRATION OF THE TWO TRUTHS

The first is a political, ideological, historical factor. The theological aspect was mentioned by Professor Hiltner, but there is also a secular non-theological aspect to it. Some years ago, N. Pastore did a study on the views of scientists on human nature. He compared the political views of twenty-four psychologists, sociologists, and biologists on their attitudes toward the nature/nurture issue. Among the twelve scientists who thought of themselves as liberals or radicals, eleven were environmentalists and only one was hereditarian. And among the twelve conservatives eleven were hereditarians and one was an environmentalist. They split right down the middle. The ideological resistance to the human-nature perspective derives from the hidden assumption that biology is destiny, an assumption shared by conservatives and liberals from Hobbes and Machiavelli to the present day. The belief in a fixed and generally unsavory human nature is usually accompanied by politically conservative beliefs. Professor Hiltner explained how such a view, which emphasizes the negative aspect of our human nature, has also had an important place in theological history.

Conservatism in politics and the assumption of a fixed human nature go hand in hand for the simple reason that while environmentalists seem to imply the possibility of radical social change, a fixed human nature seems to imply that any large-scale change will come to grief because "you can't change human nature." The presumption is that the existing social order accurately reflects human nature. The historic response of liberals to this conservative assumption, from a logical point of view, has been a curious one. They never have really quarreled with the original highly dubious premise that you can't change human nature. Instead, they have finessed the problem by introducing the even more dubious premise that there is no such thing as human nature, that the human person is infinitely plastic and malleable, a creature of circumstance, society, and culture.

The classic statement of the extreme environmentalist

position was made by the behaviorist, J. B. Watson, in the 1920s. He said, "Give me a dozen healthy infants, well-formed, and my own specified world to bring them up in, and I will guarantee to take any one at random and train him to become any kind of specialist I might select—doctor, lawyer, artist, merchant, chief, and yes, even beggarman and thief—regardless of the talent, penchant, tendencies, abilities, location, and the race of his ancestors." In our own period, an identical point of view is held by Watson's spiritual descendant, B. F. Skinner. These political and ideological considerations, then, constitute one line of resistance to the integration of the two perspectives.

A second line of resistance is somewhat more parochial. It comes from within the history of the social sciences themselves. In the early years of the century, up to about the mid-1920s, the human-nature perspective prevailed in American social psychology and sociology as represented by the instinct theory of W. McDougall. In the 1920s and the 1930s this perspective received a number of blows from which it never recovered. From within psychology one attack was launched by J.B. Watson and F. Allport, who were caught up in the logical positivist ideology of the day and were beguiled by the laboratory method of the physical sciences which emphasized the overt and the measurable. The more telling blow came from within anthropology, where anthropologists, on the basis of field studies of different cultures, had reached the conclusion that the prevailing concept of instinct could not be documented. Instead of finding a commonality of behavior, they found such a rich diversity of behavior, rituals, dreams, values, myths, and attitudes that they came to the conclusion that instinct and fixed behavioral patterns based on instinct were unfounded.

The third and most profound resistance, and in my view the one from which the other two originate, is the presence of certain logical difficulties in the formulation of the nature/nurture problem and the assumption that by human nature we mean fixed patterns of behavior such as one finds in certain species: i.e., beavers build dams, birds build nests, rats hoard, and so forth. Bertrand Russell once remarked that the metaphysics of the Stone Age is built into the very structure of our grammar and language. So it is with those famous dictums that have long plagued the history of philosophy, such as the nature/nurture problem. Certain logical and metaphysical presuppositions lead us to categorize biological influences as nature and environmental influences as nurture. These are conceived as separate categories which, in some mysterious fashion, interact to form the human person. This practice of assuming unwittingly that the referents of high abstractions such as nature and nurture exist in time/space, much as in the same way a rock or flower exists, exemplify what Whitehead called the fallacy of misplaced concreteness. All too often social scientists have conceived of human nature as if in some peculiar, disembodied sense it had to be a thing, a measurable act of behavior, a quantum of instinctive energy, or some other object reducible to physical/chemical categories. And, of course, they have failed to find it.

Because of my own indoctrination in logical positivism many years ago, it took me a long time to come to grips with the philosophical puzzle that I attempted to analyze with William Barrett in *Ego and Instinct*. I will spare you even a summary of that long argument, except to say that I believe that it lies at the heart of the conceptual confusion that causes so many intelligent and insightful persons to assume that, if they are going to give the environment its just due, they must in practice disregard the overwhelming evidence of seventy million years of evolution. Armed with a one-sided and faulty conceptual scheme, they correctly perceive the immense diversity of human behavior across cultures, but they fail to account for the even more impressive uniformity in human development, such as the fact that all humans speak some language or other. All languages contain elaborate classification systems and struc-tures of grammar that can be decoded by outsiders who do not speak the language. All cultures exhibit status systems, courtship processes, gift-giving customs, hero myths. In all societies, perceptions of the sacred and the profane can be found embodied in belief systems in which the supernatural, myths, and legends play a role. Dance ceremonials are universal, as are genital covering, imitation and socialization processes, property rules, child nurture rules, and incest rules. The list can be expanded indefinitely.

The main point I wish to make is that the conceptual task of integrating the two truths—that the human person exists only in culture, and the companion truth that he has also a specific, wide-ranging, and universal human nature—is not an im-possibly difficult task, if one can first confront the resistances that stand in the way of dealing with it.

Having struggled with these resistances, I have found it

useful to take a historical point of view. Accordingly, I would like to begin with American social history from the end of World War II and take us up to the present, using a sociological model to interpret that segment of our history. Then, against this historical background, I would like to expose certain inadequacies in the sociological model. Finally, I will attempt to relate my conclusions to those of Professor Kreyche.

POSTWAR AMERICA FROM A SOCIOLOGICAL PERSPECTIVE

The sociological tradition has been concerned with one key problem, one master question, from its inception: What is it that holds a society together? All of the great sociologists of the nineteenth and twentieth centuries have attempted to give their own answers to this question. One formula I find particularly persuasive is that given by the sociologist Emile Durkheim, who in his studies of suicide displays insight concerning the factors that might lead to a society falling apart. If you turn his perspective around you have a formula, from a Durkheimian sociological point of view, of what it is that holds a society together.

Durkheim stresses three factors. The first is the presence of a commonly held, commonly shared structure of goals and values. The second is a form of what one might call institutional legitimacy. That is to say, the institutions of the society must be held by the people they serve to have a kind of moral authority, willingly accepted, as distinct from naked power. The third factor is an unconfused sense of self held by the individuals who make up the society. Compositely, these three requirements form, in effect, a form of unwritten social contract so that one can look at any society from the point of (a) its structure of goals and instrumental values, (b) the state of the institutional legitimacy of the institutions in the society, and (c) the state of the sense of self or degree of alienation or identification of people within a society.

Looking at empirical data on American society and American attitudes and values from the end of World War II, I have arrived at the hypothesis that from the end of World War II, about 1945, to about the mid-sixties, a period of about twenty years, a strong and powerful unwritten social contract prevailed in the United States. And then in the mid-sixties, it began to change and to erode in some fairly far-reaching ways.

If we pause to look at this period from the Durkheimian perspective, the following picture emerges. The structure of goals and instrumental values in the society were, to a remarkable degree, organized around a particular meaning of success. This meaning had several elements: (1) the ownership of material possessions—automobiles, appliances, homes, money; (2) success in mobility—bettering oneself; and (3) vicarious success through one's children (even to the extent of sacrificing for them), if one could not have direct access to success.

There were a number of instrumental values that supported this conception of success rather firmly. More than 80 percent of the population subscribed to the belief that "hard work always pays off," and everyone knew what hard work and payoff meant. The family was organized around a fairly rigid division of roles. When people were asked what a "real man" was, the dominant definition of masculinity in the society was not physical strength, handsomeness, or sexual prowess, but the ability to be a good provider. The man was the provider; the woman was mother and housewife. Education was valued as "the royal road to success," and various forms of authority, patriotism, and institutional religion were widely adhered to as supportive of this system of goals and values.

With respect to institutional legitimacy, the institutions of society were held in high regard. And they validated that high regard by their performance. From 1950 to 1972, in one generation, average family income, in terms of constant 1970 dollars controlling for inflation, jumped from $5,600 in 1950 to $12,000 in 1972. We moved from a traditional two-tier society, with a segment of the population being well-to-do and the majority being poor, to a three-tier society: while the same or perhaps a larger segment of the well-to-do persisted, the proportion of those who were poor decreased steadily throughout those years up to about 1972, and the mass of the public moved into the middle-income bracket. The same pattern was found with respect to education. There were enormously large numbers of high school graduates, young people in college, and college graduates. There existed also, from the perspective of today, remarkable economic stability. The level of inflation up to 1966 was approximately 1.6 percent, when in that year the escalation in the Vietnam War led to a 3.6 percent increase in prices. On the economic security front, unemployment was gradually brought down from 6 percent to 4 percent. Goods were available. The government reinforced this particular conception of success through its public policies, such as the highway fund which encouraged development of the automobile, home subsidies to veterans which created the suburbs and shopping centers, and various forms of support for education.

It is useful to think in terms of three types of institutional legitimacy: *ideological legitimacy* where people express their support or lack of support for the basic economic and political rules of the day—free enterprise, private property, and the like; *functional legitimacy* where people believe that the institutions are doing their rudimentary jobs: the trains are running on time, garbage collectors are picking up the garbage, police are patrolling the streets, companies are making products and services; and a *moral legitimacy* where institutions are believed to be not only serving their own ends but also serving the public good. In this period from 1945 to the mid-sixties, all the institutions of the society got high marks in all three categories of institutional legitimacy.

With respect to the third Durkheimian criterion—a conception of self—there was in those years a clear-cut conception of self in the sense of social role. People had a fairly distinct identity of self as husband, provider, wife, mother, homeowner, professional, citizen, liberal, suburbanite, anti-communist. The social roles were clear cut; people were able to identify with them. There was also, at that time, a conception of professionalism that young people coming out of college shared. It was a view that, in order to have all of life's satisfactions, one could split one's life down the middle: one would render unto Caesar that which is Caesar's, give to the corporation or the employer one's mind and energies during the day, but one's real life would be lived at night and on weekends at home in the suburbs with spouse and a lot of babies and close friends. In that way, one could build a life that was meaningful, and most people were willing to accept that particular compromise.

With the 1960s there came the beginnings of the emergence of new goals and new values, a changed sense of institutional legitimacy, and a changing conception of self. To me, tracing the trajectory of this change, documenting it, describing it, understanding it, is a most interesting assignment. There is a lot of data on it, and I wish I could be less schematic, but I am going to have to touch on it very briefly.

As everyone knows, the new values and new perspectives emerged first on campus. I think it is important to recall that the campus transformation of values in the 1960s did not start off essentially as a political movement. It started as a transvaluation of conventional social values in California. In the early sixties, it became politicized with the Vietnam War. With the ending of the war, it has once again gone back to its original thrust, which is that of changing social values. Looking at that student revolution now in retrospect, and asking what the essence of these new values prompted by the students were, it seems to me that they come down to this: first, there was a radical questioning of the basic goal of the generation of the 1950s, the goal of success. If to achieve that type of success meant a nose-to-the-grindstone way of life, then it was not worth it. The epochal film of that era, *The Graduate,* captures that point very well indeed. Along with the view that success, as defined in the several dimensions I mentioned earlier, was simply not worth the nose-to-the-grindstone way of life, came a questioning of the instrumental values that went along with success. The belief that hard work always pays off declined among students from the mid-sixties to the early seventies from 76 percent endorsing that point of view to 44 percent. The view that emphasized money and the patriotic belief in "my country right or wrong" dropped in the same fashion from majority levels of support.

Accompanying the questioning of the values associated with success, came a rejection of restraints and authority of almost all forms in the direction of what sociologists call deauthorization. The authority of institutions and all forms of constraint were challenged. A new morality emerged, built not only around new sexual mores but also around a conception of self in terms of self-actualization, usually defined in contrast to role obligations to others. A new view of professionalism appeared on campus which held that it was not necessary to split one's life down the middle and derive one's economic benefits from work and one's emotional benefits solely from the family. Why should not work also be a focus of self-fulfillment and challenge? Why is it necessary to accept an artificial split in one's life?

At the same time a new sense of entitlement emerged: the notion that one has a right, as a person and a citizen, to certain things in the society hitherto regarded as a matter of privilege, or luck, or circumstance. People began to feel that they had the right to the best medical care that money could buy whether or not they could afford it, the right to college education, the right to participate in decisions in their place of employment that affected their own jobs, and so forth. In addition, a new set of values emerged built around a new relationship to nature: the old notion of putting nature to the rack gave rise to a conception of harmony with nature. Nature was defined in a broad variety of senses, human nature as well as physical nature. Multiple meanings emerged. At one time I counted seventeen different meanings of the new naturalism.

It has been fascinating to see how these various themes that emerged on campus have diffused themselves throughout the society. There appears to be a very specific transmission pathway in our society for new values, at least in this era. The pathway goes from college youth to noncollege youth, to the upper middle class adults, to the middle and lower middle class adults, urban and then rural. A number of studies among eighteen to twenty-five year old noncollege youth show that they had reached exactly the same place in their thinking about new values that the college youth had reached about five years ago, and since they had further to go, their's was a sharper and quicker trip. It was, therefore, with remarkable speed that the new ideas relating to hard work, sexual freedom, money, the clean moral life, patriotism and religion were transmitted from the sons and daughters of the well-to-do and the well educated to the mass of young people.

Abraham Maslow was probably the spiritual father of these new values with his emphasis on a hierarchy of needs, on self-actualization coming at the peak of the hierarchy, on peak experience, on the quest for excitement, and on the inner as well as the outer quest.

As these new values moved from the campus to young people who were not in college, to their parents, and then to the society as a whole, they became transformed in certain respects. If one now contrasts the perspective of the college youth with that of the country as a whole, one finds clearly what has been added by the new values. The old definition of success with its notion of a rigid hierarchy, social mobility, consumption of products and money has not disappeared but it has been muted. To it has been added the idea of the "full rich life." People today have a somewhat different, somewhat more flexible and more pluralistic conception of success. Less emphasis is being placed on some of the old constituents of success, particularly on the idea of sacrificing for one's children and instead self-fulfillment and self-actualization are more prominent. In addition, there is more emphasis on a more casual, individual-istic life-style and a rejection of functionalism, practicality, drabness. Life is invested with a quest for nostalgic romanti-cism, an openness to mysticism, and a thirst for excitement.

The human potential movement fits right in with these gropings for new values. Conveniences and short cuts are stressed as well as a more relaxed attitude toward housekeep-ing. Clothes, for example, don't have to be "whiter than white," and the lawn can go unmowed without the world coming to an end. While this is a new life-style, it is not a radical transformation but an integration of some of the new values that have become muted as they synthesize with traditional values.

Compositely, these new features add up to a rather large-scale modification of values. The change of goals has been sharpest among the adult population. With it has come a change in the status of institutions in society. This change is very strong, sharp, and clear-cut, and it is probably the most compelling indication we have that the social contract that dominated in the earlier years is undergoing some fairly far-reaching modification. Today, 80 percent of the people believe that things are going badly in the country. This is up from 37 percent two years ago. Two-thirds of the people believe that the country is in serious and deep trouble. Most people say today that they do not trust the people in power. Almost two out of five people in the country are worse off, or say they are worse off economically, than they were a year ago. The old depression psychology which had gradually abated during the sixties has flared up again with the majority of people feeling a new form of economic insecurity. Mistrust in institutions is not simply a creature of this last year or two. The University of Michigan has been measuring, through a rather sensitive scale, this notion of mistrust in government since 1958. In 1958 the level of mistrust in national government was about 12 percent—one out of eight people. A decade later in 1968 it had doubled to 27 percent—one out of four people. Five years later, in 1973, it had doubled again to 51 percent, and since that time it has gone higher. It took one decade for it to double and then half a decade for it to double again. And now mistrust in govern-ment has reached majority status.

Mistrust of other institutions, particularly big business, is even more dramatic. In 1967 business had reached the peak of its postwar reputation with seven out of ten people in the public expressing the view that business does a good job in balancing its own profitability with service to the public, which is, as it were, the basic mandate for an institution such as business. A year or two later that had slipped to about 58 percent, and then in 1970 it fell to 34 percent, and now it is at 19 percent. Confidence in business has plunged from majority to minority status. The students in the sixties had many of the same views, but they were not shared by the rest of the country. What has happened is that this doubting and questioning of institutions has spread very widely, and of course it has increased in the past two years.

With respect to the Durkheimian category of an unconfused sense of self, there has been an enormous growth of alienation, social injustice, and confusion about self. Eighty-three percent of the country express the view that "people who live by the rules do not get a fair break." The feeling of being powerless to change things has gone from 37 percent to 65 percent in about eight years. The feeling that "the people running the country don't give a damn about people like me" has increased from minority to majority status in a very short period of time.

Thus, we see signs of change along all of the three dimensions that Durkheim spoke of: the goals and values of the society have changed in some fairly far-reaching ways; the remarkably cohesive sense of institutional legitimacy has changed quite dramatically and drastically from the 1960s to the present time; and the clear-cut sense of social identity shared by people has become confused and clouded by a sense of alienation.

What we have today is, in effect, a three-pronged attack on the social contract that dominated the postwar era. One prong of the attack comes from the value transformations spearheaded by the college student movement; this is the attack from the "haves" and from the well-to-do. The affluent have questioned the social contract, not because it failed them, but precisely because they have succeeded, are ready to move on to something else and are not willing to pay what is regarded as the nose-to-the-grindstone price for the kind of success and economic well-being represented by the traditional goals of the postwar period.

The second prong of the attack comes from those who have not made it, from the people who have *not* enjoyed economic success and security. The grounds of their challenge to the social contract is not that their values have changed so much but that their views about the institutions delivering the goods have changed. Because of changing economic conditions, they question whether our institutions are going to be able to perform to the extent that they were in the past.

The third prong of the attack comes from certain epochal events like Vietnam and Watergate that have added to the erosion and undermining of confidence in our institutions.

THE SOCIOLOGICAL PERSPECTIVE CRITICIZED

Such, then, is the picture we see if we look at the country from the perspective of value change. Taking the Durkheimian point of view, we have looked at what holds a society together with the stress on goals, legitimacy, and identification. If against this historical perspective we introduce our question—Is it possible to be human and to fulfill one's human potential in a society dominated by technology and impersonal social institutions? And if so, how?—and if we add a more psychological, perhaps philosophical perspective, then I think we might perceive several fundamental flaws both in the sociological Durkheimian perspective and also substantively in the specific content of our now troubled and changing social contract.

Maybe the social contract that dominated our country for a generation worked in the past. Maybe it met the test of social stability. But it does not look so good for the future from several points of view. One is that very egalitarian American value justification for technology and impersonal social institutions advanced by Professor Kreyche, namely, the egalitarianism of the institution. The merit of the discount store and the junior college is presumably that they spread the possibility of being human to large numbers of people. This is in sharp contrast to the notion that if you are wealthy it must be because you have robbed what I have. But that has been within the continental confines of the United States. We have become aware in the past several years of the fact that the United States, with 6 percent of the world's population, has been using almost 40 percent of the world's resources. Even if we wanted to go ahead with advancing our conception of success in the form of an ever-increasing material standard of living, we probably are not going to be permitted to do so by the rest of the world. If you take the United States in relationship to the rest of the world and apply the egalitarian principles that have characterized our culture, then it raises questions about that particular social contract with its emphases on an ever-increasing material standard of living as the definition of success and the social goal toward which the human person organizes his life.

Secondly, and I believe probably the fundamental flaw in the sociological perspective, and concretely in our old social contract, is that it fits people into the system rather than designs the system to fit the hunan needs of people. As our economy and our technology and our institutions are now constituted, it is as if the purpose of people—often referred to impersonally as manpower—was to meet the needs of the economy rather than have an economy designed to meet the needs of people. If you consider the organization of work in institutions, even at the managerial level people are freely uprooted and moved from place to place wherever the company needs them. Industrial engineering approaches a job breakdown into specialized compartments so that all too often the individual worker does not take responsibility for the end product, does not participate in decisions that affect his job, and is not able to develop a sense of service which is terribly important to him or her personally. In a system where our economic security means everything to us as a nation, and for which we have sacrificed a great deal, much is jeopardized when people lose jobs through no fault of their own but simply through the workings of the impersonal market mechanism. In this process many essential human needs are sacrificed. I do not want to oversimplify the problem. The twin goals of effective economic performance and quality of working life are not inherently incompatible, but their compatibility does not occur automatically. There is very little effort going forward to achieve their integration because of the assumption that there is a system and that people must in some way fit into the system without any perspective of what essential human needs are at issue.

The willingness to sacrifice everything for the sake of economic well-being and economic security, which so dominated the period after the last depression and the postwar period, has led to many other forms of insecurity that are in many ways as threatening, and perhaps more threatening to humanness. From the point of view of inherent human needs, our social contract has been lop-sided, and this has gradually become self-evident. The sociological perspective and the way our institutions are organized also assumes that goals and values are static rather than dynamic. Thoreau said more than one hundred years ago that Americans know more about making a living than about how to live. During the past decade people have been moving from the stage of making a living to the stage of experimenting with how to live. The emerging transformation of goals and values tells us something about how people all over the country in different strata and at various age levels are struggling to evolve ways of finding out how to live as distinct from simply how to make a living.

The past century produced a number of great critiques of our society. The Marxist critique is one: it argues that we suppress an essential aspect of our humanity through the capitalist organization of work. A second, the critique of the sociologist Max Weber, offers the view that we have suppressed an essential part of our humanity, the mystical and the personal element, through the process of bureaucratization. Another, the existential critique, stresses that we have suppressed an essential part of our humanity through objectification, the converting of persons to objects. Professor Kreyche cited Gabriel Marcel's view that in reducing the person to his social role, the risk of suppressing being itself is present. The Freudian critique suggests that we have suppressed an essential part of our humanity through the sacrifice of our sexuality for the sake of preserving our particular civilization. And the conservative critique, from Burke to Nisbet, has stressed that our emphasis on the individual as an autonomous unit rather than a part of a larger whole is causing us to destroy our social bonds—what Burke called the "inns and resting places of the human spirit"—those intermediary institutions, the family, the church, one's place in the community, that lie between the individual and the state.

All of these great critiques give us a profound insight into the human spirit. All stress different aspects of humanity; all point the finger at different causes for estrangement. They leave us confused and bewildered because we respond to their diversity. But please note that at a certain level of generality they all share one assumption in common: they all imply that it is bad to suppress certain aspects of being human.

But let me remind you that it is not possible to be human as an individual or as a civilization without suppressing something: to realize one set of possibilities necessarily means to suppress another. We tend to be appalled today at the idea of suppressing any possibility. When we are young many possibilities lie open before us, and we are shaped by the choices we make. These choices involve suppressing some aspects of identity, some roles, some possibilities. So it is with a civilization. Under the old social contract, we chose to suppress certain values and possibilities and to realize others, and now we are in the process of changing our minds. The transition is a very difficult one for a culture. But the issue is not suppression versus no suppression; it is the question of which possibilities are to be suppressed and which ones are to be elicited and fulfilled, and by what criteria. There can only be one sound criterion—which is some sense, however vague, of the essence of our humanity that would be damaged by the suppression of certain possibilities.

CONCLUDING GENERALIZATION

Let me now conclude by offering a response to four general questions arising from the previous analysis. To the question, "Can we be fully human in a society dominated by technology and impersonal institutions?" my response is: Yes, we can fulfill our essential human strivings and potentiality in a technological, impersonal, institution-dominated society. Such societies are harder to live in. They make it difficult to fulfill one's human possibilities. But they are not the inherent cause of our present problem.

Do our particular technologies and impersonal institutions suppress essential human needs? My response to this question is that they probably do, not because they do so inherently, but because we have gone too far in trying to fit people to the system rather than modifying the system to fit people's needs. What is at issue, therefore, is more than mere inconveniences. We can, perhaps, accept and even embrace the rough edges of a democratic and mass culture, and yet, I think, we sense a fundamental trap that lies not inherently in technology and impersonal institutions but in a particular form of subservience to technology.

With respect to the issue of the ethics of institutions, I believe there is a fatal flaw in the moral outlook of many of the people who run our major institutions. They identify so closely with the objectives of their institutions that they inhibit their own moral sensibilities. They would not do this if it were a personal issue. If the goals of our institutions are too narrow—to win an election, to make a profit, to grow bigger, to produce more, to survive at all costs—and if the individual out of a sense of zealousness totally identifies with the organization, you have a situation which Reinhold Niebuhr described years ago under the title *Moral Man and Immoral Society*. Under these circumstances, we must strive to humanize the goals of the institutions.

Finally there is the question of human relationships themselves. To me, Maslow, with his emphasis on self-actualization, is in certain respects a false prophet for our present crisis. I believe that our survival as an endangered species depends more on an ethic of concern for others than on an extension of a "do-your-own-thing" form of individualism. The focus on self-realization, without any suppression, is the false premise that lies at the heart of the human potential movement. I believe we have to strengthen our concern for others and suppress some of our narcissistic self-concern. We must invent new institutions which will create new "inns and resting places for the human spirit," and which will interpose themselves between us and these larger institutions, even at the expense of suppressing some possibilities of purely individual self-fulfillment.

Let me stress, however, that this concern need not be a return to the old and sometimes rather cold emphasis on duty to others of the Protestant ethic: duty is forcing yourself to do something you do not want to do. I am talking about concern for others that flows from an inherent aspect of human nature that should not be suppressed but is probably innate and capable of being developed. In order to name what this characteristic is, one has to go outside our own culture. The Japanese have a word, *Amae*. To Japanese observers of other cultures and other scenes, it seems inconceivable that one could describe a culture, a society, a human relationship, without a term that corresponds to the notion of *Amae*. What is this notion of *Amae*? It is the kind of basic trust and confidence in a relationship that a child will often have for a parent where he or she feels so fully accepted that he feels free to presume on the other's good will without threatening the relationship. The closest term we have is "dependence," which really fails to capture the meaning. We need this quality of *Amae*. I believe that it is possible for us to have a society in which technology and discount stores and junior colleges and *Amae* coexist, but we do not have one now. It is going to take a lot of faith and hope and work to build it.

CHAPTER SIX

Ideology and Utopia as Cultural Imagination

Paul Ricoeur

The purpose of this paper is to put the two phenomena of ideology and utopia within a single conceptual framework which I will designate as a theory of cultural imagination. From this connection under this merely formal title, I expect two things: first, a better understanding of the ambiguity which they both have in common to the extent that each of them covers a set of expressions ranging from wholesome to pathological forms, from distorting to constitutive roles; second, a better grasp of their complementarity in a system of social action. In other words, my contention is that the polarity between ideology and utopia and the polarity within each of them may be ascribed to some structural traits of cultural imagination.

The polarity between ideology and utopia has been scarcely taken as a theme of inquiry since the time when Karl Mannheim wrote his seminal work *Ideologie und Utopie* in 1929.[1] Today we have, on the one hand, a critique of ideologies stemming from the Marxist and post-Marxist tradition and expanded by the Frankfurt school, and, on the other hand, a history of utopias, sometimes a sociology of utopia, but with little connection to the so-called *Ideologiekritik*. Yet Karl Mannheim had paved the way for a joint treatment of both ideology and utopia by looking at them as deviant attitudes toward social reality. This criterion of non-congruence, or discrepancy presupposes that individuals as well as collective entities may be related to social reality not only in the mode of a participation without distance, but also in a mode of non-congruence which may assume various forms. This presupposition is precisely that of a social or cultural imagination operating in many ways, including both constructive and destructive ones. It may be a fruitful hypothesis that the polarity of ideology and utopia has to do with different figures of non-congruence, typical of social imagination. Moreover, it is quite possible that the positive side of the one and the positive side of the other are in the same complementary relation as the negative and pathological side of the one is to the negative and pathological side of the other.

But before being able to say something about this over-arching complementarity between two phenomena which are themselves two-sided, let us speak of each phenomenon separately in order to discover the place of the one on the borderline of the other. I shall start from the pole of ideology.

In this section devoted to the phenomenon of ideology, I propose that we start from the evaluative concept of ideology, i.e., the pejorative concept in which ideology is understood as concealment and distortion. Our task will be to inquire into the presuppositions by means of which the pejorative concept of ideology makes sense. This kind of regressive procedure will lead us from the surface layer of the phenomenon to its depth structure, a procedure not intended to refute the initial concept, but to establish it on a sounder basis than the polemical claim to which it first gives expression.

I borrow this initial concept of ideology from Marx's *German Ideology*.[2] The choice of this starting point has a twofold advantage. On the one hand, it provides us with a concept of ideology which is not yet opposed to an alleged Marxist science (which is still to be written), but to the concept of the real living individuals under definite material conditions. Therefore we are not yet trapped by the insoluble problem of science versus ideology. On the other hand, *The German Ideology* is already a Marxist text which breaks with the idealistic philosophy of the young Hegelians who put "consciousness," "self-consciousness," "Man," "species-being" and "the Unique" at the root of their anthropology. A new anthropology has emerged for which reality means praxis, i.e., the activity of human individuals submitted to circumstances which are felt as compulsory and seen as powers foreign to their will.

It is against this background that ideology is defined as the sphere of representations, ideas, and conceptions versus the sphere of actual production, as the imaginary versus the real, as the way individuals "may appear *(erscheinen)* in their own or other people's representation *(Vorstellung)*," versus the way "they really *(wirklich)* are, i.e., as they operate *(wirken)*, produce materially, and hence as they work under definite material limits, presuppositions, and conditions independent of their will."[3]

The first trait of ideology therefore is this gap between the unactual representations in general (religious, political, juridical, ethical, aesthetical, etc.) and the actuality of the life-process. This first trait leads immediately to the next one: the dependence of what is less actual on what is more actual. "Life is not determined by consciousness, but consciousness by life."[4] Here we are not far from the idea that in ideology we find only "reflexes and echoes of this life-process,"[5] which implies in turn that only the practical processes of life have a history. Ideology has no history of its own, even no history at all. We may now shift easily to the decisive trait. Ideology appears as the inverted image of reality. "If in all ideology men and their circumstances appear upside-down as in a *camera obscura*, the phenomenon arises just as much from their historical life-process as the inversion of objects on the retina does from their physical life-process."[6]

This metaphor of the inverted image will provide the guideline for our inquiry. What is at stake here is not the empirical accuracy of the descriptive arguments offered by Marx, but the meaningfulness or intelligibility of the concept of ideology as an inverted image of reality.

In the *Manuscripts of 1844,*[7] an interpretation was given which relies basically on Feuerbach's notion of "estrangement," conceived as the inversion of the process of "objectification." This is the process by which man's consciousness generates its own existence by actualizing itself in some external entity or entities. Through "estrangement" the result of this radical production becomes an external power to which man becomes enslaved. Indeed, this schema of estrangement as the inversion of the process of self-objectification is no longer applied by Marx to the religious sphere as in Feuerbach, but to the sphere of labor and private property. It is labor which is estranged under the power of private property. But labor is still conceived in metaphysical terms according to the paradigm of objectification, of becoming an object in order to become oneself.

With *The German Ideology* the concept of the division of labor tends to replace that of estrangement or alienation, or at least to fill it with a more concrete content. The fragmentation of human activity becomes the equivalent of what had been called estrangement. The division of labor "offers us the first example of how as long as man remains in natural society, that is, as long as a cleavage exists between the particular and the common interest, as long, therefore, as activity is not voluntarily but naturally divided, man's own deed becomes an alien power opposed to him which enslaves him instead of being controlled by him."[8] Within this new framework ideology appears as a particular case of the division of labor. "Division of labor only becomes truly such from the moment when a division of material and mental labor appears. (The first form of ideologists, priests, is concurrent.) From this moment onwards consciousness *can* really flatter itself that it is something other than consciousness of existing practice, that it *really* represents something without representing something real; from now on consciousness is in a position to emancipate itself from the world and to proceed to the formulation of 'pure' theory, theology, philosophy, ethics, etc."[9]

So the metaphor of the inverted image refers at least to an initial phenomenon, the division of labor, the history of which may be empirically stated.

But, if the division of labor partially explains the tendency of conscious representations to become autonomous, it does not explain their tendency to become illusory. Of course, a mode of thought which would not be autonomous as regards its basis in practical life would have no chance of becoming distorted. Marx has a remark about this non-autonomous, non-distorted mode of thought which he very properly calls the "language of real life."[10] "Conceiving, thinking, the mental intercourse of men, appear at this stage as the direct efflux *(Ausfluss)* of their material behavior."[11] It is on this "language of real life" that a "real, positive science" has to be grafted, a science which would be no longer an empty "representation," but the actual depiction or presentation *(Darstellung)* of the practical activity.[12]

Division of labor therefore does not explain either the initial stage, that of the language of real life, which will later provide us with the basic concept of ideology taken in the sense of Clifford Geertz's concept of symbolic action, or the final stage, that of an autonomy of the representational world becoming an inverted image of real practical life.

Let us set aside the problem raised by this initial stage which Marx refers to as the language of life and focus on the effects of the seclusion of the intellectual process from its basis in practical life. How does autonomy generate illusion?

The gap between mere autonomy and distortion is partially filled by the insertion of the concepts of class and ruling class between the concept of the division of labor and that of ideology. *The German Ideology* explains the metaphor in the following way: the ruling class establishes its power by concealing the ideas which express its interest between the screen of idealistic thoughts. "The ideas of the ruling class are in every epoch the ruling ideas, i.e., the class which is the ruling *material* force of society, is at the same time its ruling *intellectual* force . . . The ruling ideas are nothing more than the ideal expression of the dominant material relationships, the dominant material relationships grasped as ideas; hence of the relationships which make the one class the ruling one, therefore, the ideas of its dominance."[13]

These concepts of "ruling class" and "ruling ideas" are so decisive that after they have been introduced the nuclear concept of the division of labor itself has to be referred to the class structure.

> The division of labor . . . manifests itself also in the ruling class as the division of mental and material labor, so that inside this class one part appears as the thinkers of the class (its active, conceptive ideologists, who make the perfecting of the illusion of the class about itself their chief source of livelihood), while the others' attitude to these ideas and illusions is more passive and receptive, because they are in reality the active members of this class and have less time to make up illusions and ideas about themselves.[14]

If we accept Marx's assumption that "the ideas of a ruling class are in every epoch the ruling ideas," it remains to be explained how "dominant material relationships" become "ruling ideas." Marx says that ruling ideas are the "ideal expression" of these relationships, but two difficulties are implied here. I shall put aside the first one which concerns the notion of an idea "expressing" a process rooted in practical life and admittedly prior to consciousness, representations, and ideas. What is at stake here is the very dichotomy between real and imaginary evoked at the beginning of our analysis of the concept of ideology. Since Marx himself suggests by his allusion to the "language of life" that there must be a place or a stage in which praxis itself implies some symbolic mediation, I shall return to this point later to show that the concept of distortion only makes sense if it applies to a previous process of symbolization constitutive of action as such. This will provide us with the final concept of ideology.

Let us rather focus on the second difficulty implied by the

statement that the ideas of the ruling class are in every epoch
the ruling ideas because the ruling ideas are held to be the ideal
expression of the dominant material relationships. This
difficulty concerns the process of idealization by which an
expression becomes a ruling idea. What is an ideal expression?
Marx explains this idealization in the following way.

> For each new class which puts itself in the place of one ruling
> before it is compelled, merely in order to carry through its aim, to
> represent its interest as the common interest of all the members of
> society, that is, expressed in ideal form: it has to give its ideas the
> form of universality, and represent them as the only rational,
> universally valid ones.[15]

According to this explanation the necessity to represent a
particular interest as general is the key to the process of
idealization. The metaphor of the inverted image borrowed
from the experience of the *camera obscura* and already
extended to the image on the retina loses much of its enigmatic
obscurity when it is related to the substitution of the rule of
certain ideas for the rule of a certain class. The inverted image
is "this whole semblance, that the rule of a certain class is only
the rule of certain ideas."[16]

But has the enigma of the inverted image become completely
transparent? This can be questioned. How can a particular
interest be represented as general? This role of representation
as the concealment of the particularity of interest under a claim
to generality is more the name of a problem than that of a
solution. Is there only one way to proceed to this concealment?
Are all the cultural products of the bourgeoisie in the
seventeenth and eighteenth centuries, for example, equally
such false representations? How can we account for their
immense variety? Can they be reduced to a unique ideological
field? If so, how does the ideological field of an epoch, taken as a
unique network, refer to its real basis, i.e., the system of
interests of the so-called ruling class?

Orthodox Marxism has attempted to solve these paradoxes
by assuming that a causal relation holds between the economic
basis and the ideological superstructure. This causal
relationship is such that, on the one hand, the mode of
production determines in the last instance the superstructure
while, on the other hand, the superstructure enjoys a relative
autonomy and a specific effectivity. Production is the determi-
nant factor, but only in the last instance. Engels will refine this
formula in his well-known letter to Bloch of September 21,
1890.[17]

Unfortunately this formula only gives us the two ends of the chain, somewhat like those formulas of theology which attempt to tie together divine predestination and human free will. In fact, nobody is able to discover what goes on between them. Why? Because the problem is insoluble so long as it is put within the framework of causal relationships between structures, as we do when we speak of relative effectivity and of determination in the last instance. Before being able to speak of relative or ultimate effectivity, we must inquire whether the question has been posed in terms which make sense. I should like to suggest that Marx himself had opened a more fruitful path when he declared that "each new class which puts itself in the place of one ruling before it is compelled merely in order to carry through its aim, to represent its interests as the common interest of every member of society. . . ."[18] According to this formulation the relation between the interest and its ideal expression cannot be put within the framework of causation, but requires something like a relation of motivation. What is at stake here is a process of legitimation, of justification, described by Marx as a necessity to represent a particular interest as general, as the only rational, universally valid one.

But besides the fundamental obscurity of the notion of an interest "expressed" in ideas, the process which gives ideas the form of universality has also to be explained. This cunning of interest, substituted for the Hegelian cunning of reason, remains enigmatic. On the one hand, it presupposes that the notions of rationality and universal validity make sense by themselves, besides and before their fraudulent capture by the ruse of interest. On the other hand, this capture itself presupposes that domination cannot succeed without the acceptance of the arguments offered to legitimize the claims of the ruling class. This connection between domination and legitimation constitutes in my opinion one of the two unsolved enigmas of the Marxist concept of ideology, the second being more radical in that it concerns the fundamental tie between an interest and its alleged expression.

Both difficulties exceed the capabilities of Marxist thought. The first one, the connection between the ruling class and the ruling ideas, is only a particular case of the larger problem of the relation between domination and legitimation. To say this is not to diminish the merit of Marx. He has delineated a fundamental source of ideology by connecting it to the central structure of domination embodied in the class structure of society. But it is not certain that the class structure and its corollary notion of a ruling class exhaust the phenomenon of domination. It is quite possible that both the notions of class and ruling class display only one side or one aspect of the problem of domination.

It has been the great merit of Max Weber to have approached the problem of domination as a specific problem. In *Wirtschaft und Gesellschaft*[19] he first discusses the typology of order in corporate groups as a problem of its own. Then he refers the functioning of power (*Macht*), or domination or imperative control (*Herrschaft*), to this typology of order. Only then does he introduce the notion of political power as one kind of "imperatively coordinated corporate group." The state is the compulsory political association implying continuous organization. This kind of organization "will be called a state if and only insofar as its administrative staff successfully upholds a claim to the *monopoly* of the *legitimate* use of physical force in the enforcement of its order."[20] It is within this broad framework that the problem of the basis of legitimacy may be raised. It is raised in the following terms: "It is an induction from experience that no system of authority voluntarily limits itself to the appeal to material or effectual or ideal motives as a basis for guaranteeing its continuance. In addition every such system attempts to establish and to cultivate the belief in its 'legitimacy.' "[21]

Therefore the ground on which this problem makes sense is that of human action as having motives. The belief in the existence of a legitimate order relies on this assumption. The problem of validity cannot be raised in other terms than those of the motivation of meaningful action. It presupposes that the types of authority can be classified according to the kind of claims to legitimacy typically made by each.

My contention is that the problem raised by Marx about the relation between the ruling class and ruling ideas is capable, if not of a complete solution, then at least of a rational treatment when seen in this fashion. The question now is that of the relation between a claim to legitimacy and a belief in legitimacy, a claim raised by the authority and a belief conceded by the individuals. As is well known, Max Weber considered three types of legitimate authority according to the basis alleged for the validity of these claims: rational, traditional, and charismatic grounds. This typology is not our problem here. Our problem is that of correctly "placing" ideology in this process.

I wonder whether the function of ideology here is not to fill up what we could call a credibility gap. By this I mean the unavoidable excess of the claim over against the belief. In this sense, I should be tempted to speak of the attempt to fill this gap as a case of overvalue, to borrow a term that Marx used to characterize the surplus of value provided by labor and diverted by the owners of capital. Is it not the case that any authority always claims more than what we can offer in terms of belief? If this is the case, then could we not say that the main function of a system of ideology is to reinforce the belief in the legitimacy of the given systems of authority in such a way that it meets the claim to legitimacy? Ideology would be the system of justification capable of filling up the gap of political overvalue.

With this function of justification the aspect of distortion becomes more understandable. The relation between claim and belief which is described in terms of overvalue is the place *par excellence* of dissimulation and distortion. No system of legitimacy is completely transparent. The process which Marx describes as giving "ideas the form of universality, and presenting them as the only rational, universally valid ones," makes sense as the kind of distortion required by the claim to legitimacy. But it only makes sense under several conditions. First, that an interest asserts itself at the level of power or authority. Second, that authority makes itself acceptable at the level of a claim to legitimacy and not only at the level of sheer application of force. Third, that rationality is understood for its own sake as the general horizon of understanding and mutual recognition before being unduly diverted for the sake of a ruling group, be it a class or any other dominant group. Whatever may be the complex relationships between interest, authority, legitimacy, belief, and ideology, these factors work and make sense within a system of motivation, not of causation.

We are not ready to address the most difficult problem of all. The preceding discussion moves within the sphere of ideas. Ideology gives ideas the form of universality. But what about the assumed relationship between interest and idea? Are we not too easily satisfied with the assumption that interests "express" themselves through ideas? The initial dichotomy Marx imposed on the whole problem—I mean the distinction between what people are and do, on the one hand, and how they appear in their own or other people's imagination, on the other—at least has the advantage of transforming a triviality into a paradox: if praxis and representation move on different planes, how can the latter express the former? This question is raised in its most radical way by Marx's phrase, "the language of real life," and by the claim grafted on this assumption that a real and positive science is possible which would be the depiction or presentation (*Darstellung*) of practical life.

Here Clifford Geertz may be helpful when he relates all the distorting functions of ideology to a more basic function, that of mediating and integrating human action at its public level. In his seminal article, "Ideology as a Cultural System," reprinted in *The Interpretation of Cultures*,[22] he shows very clearly that the main available theories of ideologies, Marxist and non-Marxist, fail to give a meaningful account of the concept of expression in such phrases as "the expression of interests or of conflicts in the sphere of ideas." He forcefully demonstrates that an interest theory as well as a strain theory fail to show how ideologies transform sentiment into significance and make it socially available. In both theories the diagnostic is sound, but the explanation is deficient. The reason is that in both cases the autonomous process of symbolic formulation has been overlooked.

To fill this lacuna, he suggests that we apply the concept of symbolic action advocated by Kenneth Burke in his *Philosophy of Literary Forms: Studies in Symbolic Action*[23] to the theory of ideologies. What these theories of ideology fail to understand is that action in its most elementary forms is already mediated and articulated by symbolic systems. If this is the case, the explanation of action has to be itself mediated by an interpretation of its ruling symbols. Without recourse to the ultimate layer of symbolic action, of action symbolically articulated, ideology has to appear as the intellectual depravity that its opponents aim to unmask. But this therapeutic enterprise is itself senseless if it is incapable of relating the mask to the face. This cannot be done as long as the rhetorical force of the surface ideology is not related to that of the depth layer of symbolic systems which constitute and integrate the social phenomenon as such.

How shall we interpret this integrative function? Geertz is right, I think, when he suggests transferring some of the methods and results of literary criticism to the field of the sociology of culture and treating ideology as a kind of figurative language. "With no notion of how metaphor, analogy, irony, ambiguity, pun, paradox, hyperbole, rhythm, and all the other elements of what we lamely call style operate . . . in casting personal attitudes into public form,"[24] it is impossible to construe the import of ideological assertions.

The advantage of this connection between tropology and ideology is that it helps us solve the problem which is concealed more than delineated by the phrase, "the expression of interests in ideas." If the rhetoric of ideology proceeds like, say, that of metaphor, then the relation between ideology and its so-called real basis may be compared to the relation of reference which a metaphorical utterance entertains with the situation it redescribes. When Marx says that the ruling class imposes its ideas as the ruling ideas by representing them as ideal and universal, does not this device have some affinity with the hyperbole described by rhetoricians?

If this comparison between ideology and rhetorical devices works and holds, the decisive conclusion to draw is this: under the layer of distorting representation we find the layer of the systems of legitimation meeting the claim to legitimacy of the given system of authority. But under these systems of legitimation we discover the symbolic systems constitutive of action itself. As Geertz says, they provide a template or blueprint for the organization of social and psychological processes, as genetic systems provide such a template or blueprint for the organization of organic processes.

A first corollary of this statement is that the initial opposition between real active life process and distorting representations is as such meaningless if distortion is not a pathological process grafted on the structure of action symbolically articulated. If action is not symbolic from the very beginning, then no magic will be able to draw an illusion from an interest.

A second corollary is still more important because it will provide us with a transition to the problem of utopia. At its three levels, distortion, legitimation, symbolization, the ideology has one fundamental function: to pattern, to consolidate, to provide order to the course of action. Whether it preserves the power of a class, or insures the duration of a system of authority, or patterns the stable functioning of a community, ideology has a function of conservation in both a good and a bad sense of the word. It preserves, it conserves, in the sense of making firm the human order that could be shattered by natural or historical forces, by external or internal disturbances. All the pathology of ideology proceeds from this "conservative" role of ideology.

The shadow of the forces capable of shattering a given order is already the shadow of an alternative order that could be opposed to the given order. It is the function of utopia to give the force of discourse to this possibility. What hinders us from recognizing this connection between ideology and utopia is precisely what appears at first glance to be utopia's place in discourse. In its strict sense, utopia is a literary genre.

Thomas More coined the term in 1516 as a title for his famous book, *Utopia*, and the word means the island which is nowhere. "It is a place which has no place." As a genre, utopia has a literary existence. It is a way of writing. But this literary criterion may prevent us from perceiving the complementarity and, in general, the subtle relationships between ideology and utopia. Ideology has no literary existence, since it has no knowledge of itself, whereas utopia asserts itself as utopia and knows itself as utopian. This is why utopia may be claimed by its author, whereas I know of no author who would claim that what he is doing is ideology, except for the French "ideologists" of the eighteenth century. But that was before Napoleon made their name infamous. We may name authors of utopias, but we are unable to ascribe ideologies to specific authors.

Moreover, we are related to ideology by a process of unmasking which implies that we do not share in what Marx called the illusions of the epoch, but we may read utopias without calculating or committing ourselves to the probability of their projects.

In order to initiate a parallelism between utopia and ideology, we have to proceed from the literary genre to the "utopian mode," to use a distinction borrowed from Raymond Ruyer in his *L'Utopie et les utopies*.[25] This shift implies that we forget the literary structure of utopia and also that we overcome the specific contents of proposed utopias. As long as we remain at the level of their thematic content, we will be disappointed to discover that in spite of the permanence of certain themes such as the status of the family, the consumption of goods, the appropriation of things, the organization of political life, and the role of religion, each of these topics is treated in such a variety of ways as to imply the most contradictory proposals for changing society. This paradox provides us with a clue for interpreting the utopian mode in terms of a theory of imagination rather than emphasizing its content.

The utopian mode is to the existence of society what invention is to scientific knowledge. The utopian mode may be defined as the imaginary project of another kind of society, of another reality, another world. Imagination is here con-stitutive in an inventive rather, than an integrative manner, to use an expression of Henri Desroche.

If this general feature of the utopian mode holds, it is easy to understand why the search for "otherness" has no thematic unity, but instead implies the most diverse and opposed claims. Another family, another sexuality, may mean monasticism or sexual community. Another way of consuming may mean asceticism or sumptuous consumption. Another relation to property may mean direct appropriation without rules as in many "Robinsonades" or artificial accurate economic plan-ning. Another relation to government may mean self-government or authoritarian rule under a virtuous and disciplined bureaucracy. Another relation to religion may mean radical atheism or new cultic festivity. And we could make numerous additions to these variations on the theme of "otherness" in every domain of communal existence.

Another step, however, leads beyond the mode of utopia to "the spirit of utopia," to once again use Raymond Ruyer's categories. To this spirit belong the fundamental ambiguities which have been assigned to utopia and which affect its social function. We discover at this level a range of functional variations which may be paralleled with those of ideology and which sometimes intersect those functions which earlier we described as ranging from the integrative to the distorting.

At this stage of our analysis the regressive method we applied to ideology may be helpful for disentangling the ambiguities of the utopian spirit. Just as we are tempted by the Marxist tradition to interpret ideology in terms of delusion, so we may be inclined to construe the concept of utopia on the basis of its quasi-pathological expressions. But let us resist this temptation and follow a course of analysis similar to the one which we followed concerning ideology.

Let us begin from the kernel idea of "nowhere" implied by the very word "utopia" and Thomas More's description: A place which has no place, a ghost city; for a river, no water; for a prince, no people, etc. What must be emphasized is the benefit of this kind of extra-territoriality for the social function of utopia. From this "no-place" an exterior glance is cast on our reality, which suddenly looks strange, nothing more being taken for granted. The field of the possible is now opened beyond that of the actual, a field for alternative ways of living. The question therefore is whether imagination could have any constitutive role without this leap outside. Utopia is the way in which we radically rethink what is family, consumption, government, religion, etc. The fantasy of an alternative society and its topographical figuration "nowhere" works as the most formidable contestation of what is. What some, for example, call conscientization (mainly in Latin America), or what elsewhere is called cultural revolution, proceeds from the possible to the real, from fantasy to reality.

Utopia thus appears as the counterpart of the basic concept of ideology where it is understood as a function of social integration. By way of contrast, utopia appears as the function of social subversion.

Having said this, we can extend our parallelism a step further following the intermediate concept of ideology, ideology understood as a tool of legitimation applied to given systems of authority. Ultimately what is at stake in utopia is the apparent givenness of every system of authority. And our previous interpretation of the process of legitimation gives us a clue to the way in which utopia works at this level. We assumed that one of the functions, if not the main one, of ideology was to provide a kind of overvalue or surplus value to the belief in the validity of authority such that the system of power may implement its claim to legitimacy. If it is true that ideologies tend to bridge the credibility gap of every system of authority and eventually to dissimulate it, could we not say that it is one of the functions of utopia, if not its main function, to reveal the undeclared overvalue and in that way to unmask the pretense proper to every system of legitimacy? In other words, utopias always imply alternative ways of using power, whether in family, political, economic or religious life, and in that way they call established systems of power into question.

Once again, this function may assume different forms at the level of thematic content. Another society means another power, either a more rational power, or a more ethical power, or a null power if it is claimed that power as such is ultimately bad or beyond rescue.

That the problematic of power is the kernel problem of every utopia is confirmed not only by the description of social and political fantasies of a literary kind, but also by an examination of the various attempts to actualize utopias. The prose of the utopian genre does not exhaust the utopian mode or the utopian spirit. There are (partially) realized utopias. These are, it is true, mainly micro-societies, some more permanent than others ranging from the monastery to the kibbutz or commune. But they are utopian in the sense that they constitute kinds of laboratories or miniature experiments for broader projects involving the whole of society.

If we try to find a common trait of such diverse experiments, their common concern seems to be the exploration of the possible ways of exerting power without resorting to violence. These attempts to actualize utopia testify not only to the seriousness of the utopian spirit, but also to its aptitude to address itself to the paradoxes of power.

The modern utopias of our generation provide an additional confirmation of this thesis. They are all in one way or another directed against the abstraction, the anonymity, and the reification of the bureaucratic state. Such atoms of self-management are all challenges to the bureaucratic state. Their claim for radical equality and the complete redistribution of the ways in which decisions are made implies an alternative to the present uses of power in our society.

If it is true that ideology and utopia meet at this intermediary level of the legitimation or contestation of the system of power, it becomes understandable that the pathology of utopia corresponds also to the pathology of ideology. In the same way we are able to recognize in the positive concept of ideology, ideology as conservation, the germ of its negative counterpart, the distortion of reality and the dissimulation of its own process, so we may perceive the origin of the specific pathology of utopia in its most positive functioning. Because utopia proceeds from a leap elsewhere to "nowhere," it may display disquieting traits which may easily be discerned in its literary expressions and extended to the utopian mode and the utopian spirit: a tendency to submit reality to dreams, to delineate self-contained schemas of perfection severed from the whole course of the human experience of value. This pathology has been described as "escapism," and it may develop traits which have often been compared to those of schizophrenia: a logic of all or nothing which ignores the labor of time. Hence the preference for spatial schematisms and the projections of the future in frozen models which have to be immediately perfect, as well as its lack of care for the first steps to be taken in the direction of the ideal city. Escapism is the eclipse of praxis, the denial of the logic of action which inevitably ties undesirable evils to preferred means and which forces us to choose between equally desirable but incompatible goals. To this eclipse of praxis may be referred the flight into writing and the affinity of the utopian mode for a specific literary genre, to the extent that writing becomes a substitute for acting.

At its ultimate stage the pathology of utopia conceals under its traits of futurism the nostalgia for some paradise lost, if not a regressive yearning for the maternal womb. Then utopia, which in the beginning was most candid in the public display of its aims, appears to be no less dissimulating than ideology. In this way both pathologies cumulate their symptoms in spite of the initial opposition between the integrative and the subversive function.

The time has come to account for this double dichotomy between, first, the two poles of ideology and utopia, and second, the ambiguous variations possible internally to each pole. We shall attempt to do so in terms of the imagination.

We must begin, it seems to me, by attempting first to think about both ideology and utopia as a whole in terms of their most positive, constructive and —if we may use the term—healthy or wholesome modalities. Then using Mannheim's concept of non-congruence, it will be possible to construe the integrative function of ideology and the subversive function of utopia together.

At first glance, these two phenomena are simply the inverse of each other. But if we examine them more closely we see that they dialectically imply each other. The most "conservative" ideology, I mean one which does nothing more than parrot the social order and reinforce it, is ideology only because of the gap implied by what we might call, paraphrasing Freud, the "considerations of figurability" which are attached to the social image. Conversely, utopian imagination appears as merely eccentric and erratic. But this is only a superficial view. What decenters ourselves is also what brings us back to ourselves. So we see the paradox. On the one hand, there is no movement towards full humanity which does not go beyond the given; on the other hand, elsewhere leads back to here and now. It is, as Lévinas remarks, "as if humanity were a genus which admitted at the heart of its logical place, of its extension, a total rupture; as if in going towards the fully human, we must transcend man. It is as if utopia were not the prize of some wretched wandering, but the clearing where man is revealed."[26]

This interplay of ideology and utopia appears as an interplay of the two fundamental directions of the social imagination. The first tends towards integrations, repetition, and a mirroring of the given order. The second tends to lead us astray because it is eccentric. But the one cannot work without the other. The most repetitive or reduplicative ideology, to the extent that it mediates the immediate social ties, the social-ethical substance, as Hegel would call it, introduces a gap, distance, and consequently something potentially eccentric. And as regards utopia, its most erratic forms so long as they move within "a spere directed towards the human," remain hopeless attempts to show what man fundamentally is when viewed in the clarity of utopian existence.

This is why the tension between ideology and utopia is unsurpassable. It is even often impossible to tell whether this or that mode of thought is ideological or utopian. The line seemingly can only be drawn after the fact on the basis of a criterion of success which in turn may be called into question insofar as it is built upon the pretention that whatever succeeds is warranted. But what about abortive attempts? Do they not sometimes return at a later date and sometimes obtain the success that history had previously denied them?

The same phenomenology of social imagination gives us the key to the second aspect of our problem, namely that each term of the couple ideology-utopia develops its own pathology. If imagination is a process rather than a state of being, it becomes understandable that a specific dysfunction corresponds to each direction of the imaginative process.

The dysfunctioning of ideology is called distortion and dissimulation. We have seen above how these pathological figures constitute the privileged cases of dysfunctioning which are grafted onto the integrative function of social imagination. Here let us only add that a primitive distortion or an original dissimulation are inconceivable. It is within the symbolic constitution of the social order that the dialectic of concealment and revelation arises. The reflective function of ideology can only be understood as arising from this ambiguous dialectic which already contains all the traits of non-congruence. It follows that the tie denounced by Marxism between the process of dissimulation and the interests of a class only constitutes a partial phenomenon. Any "superstructure" may function ideologically, be it science and technology or religion and philosophical idealism.

The dysfunctioning of utopia must also be understood as arising from the pathology of the social imagination. Utopia tends towards schizophrenia just as ideology tends towards dissimulation and distortion. This pathology is rooted in the eccentric function of utopia. It develops almost as a caricature of the ambiguity of a phenomenon which oscillates between fantasy and creativity, between flight and return. "Nowhere" may or may not refer to the "here and now." But who knows whether such and such an erratic mode of existence may not prophesy the man to come? Who even knows if a certain degree of individual pathology is not the condition of social change, at least to the extent that such pathology brings to light the sclerosis of dead institutions? To put it more paradoxically, who knows whether the illness is not at the same time a part of the required therapy?

These troubling questions at least have the advantage of directing our regard towards one irreducible trait of social imagination, namely, that we only attain it across and through the figures of false consciousness. We only take possession of the creative power of imagination through a relation to such figures of false consciousness as ideology and utopia. It is as though we have to call upon the "healthy" function of ideology to cure the madness of utopia and as though the critique of ideologies can only be carried out by a conscience capable of regarding itself from the point of view of "nowhere."

NOTES

1.   Karl Mannheim, *Ideology and Utopia: An Introduction to the Sociology of Knowledge* (New York: Harcourt, Brace, Jovanovich, 1955).

2.   Karl Marx and Fredrich Engels, *The German Ideology*, ed. C.J. Arthur (New York: International Publishers, 1970).

3.   *Ibid.*, pp. 46-47.

4.   *Ibid.*, p. 47.

5.   *Ibid.*

6.   *Ibid.*

7.   Karl Marx, *The Economic and Philosophic Manuscripts of 1844*, ed. Dirk J. Struik (New York: International Publishers, 1964).

8.   *The German Ideology*, p. 53.

9.   *Ibid.*, pp. 51-52, emphasis in original text.

10.   *Ibid.*, p. 47.

11.   *Ibid.*

12.   *Ibid.*

13.   *Ibid.*, p. 64, emphasis in original text.

14.   *Ibid.*, p. 65.

15.   *Ibid.*, pp. 65-66.

16.   *Ibid.*, p. 66.

17.   Quoted in Louis Althusser, *For Marx,* trans. Ben Brewster (New York: Vintage Books, 1970), pp. 111-12.

18.   See note 15.

19.   See Max Weber, *The Theory of Social and Economic Organization,* trans. A. M. Henderson and Talcott Parsons (New York: Oxford University Press, 1947).

20.   *Ibid.,* p. 154, emphasis in original.

21.   *Ibid.,* p. 325.

22.   Clifford Geertz, *The Interpretation of Cultures* (New York: Basic Books, 1973), pp. 193-233.

23.   Kenneth Burke, *Philosophy of Literay Forms: Studies in Symbolic Action* (Berkeley: University of California Press, 1974).

24.   Geertz, *The Interpretation of Cultures,* p. 209.

25.   Paris: Presses Universitaires de France, 1950.

26.   Emmanuel Lévinas, "L'Etre et l'autre. A propos de Paul Celan," in *Sens et Existence, en hommage à Paul Ricoeur,* ed. G. B. Madison (Paris: Editions du Seuil, 1975), p. 28.

CHAPTER SEVEN

Humanness in Neo-Vedāntism

Troy Organ

The human is central in Hinduism. According to an early hymn of the *Ṛg Veda* the cosmos was created from Puruṣa, the primordial man.

> The moon was gendered from his mind, and from his eye
> the sun had birth;
> Indra and Agni from his mouth were born, and Vayu from
> his breath.
> Forth from his navel came mid-air; the sky was fashioned
> from his head;
> Earth from his feet, and from his ear the regions. Thus
> they formed the worlds.[1]

The *Mahābhārata*, the great epic of India, opens with a benedictory petition to Nārāyaṇa (Man the Deity) and to Nara (The Original Man) and near to the end announces, "I disclose to you a great mystery. There is no status that is superior to that of humanity."[2] A modern Indian philosopher has characterized Indian thought as "Humanism with a bias towards spirituality." He adds, "We may say, in short, that Indian philosophy is a running commentary on the text, 'Thanks that I am a man.' "[3] Gandhi said, "The supreme consideration is man."[4] Aurobindo reminds us that in Hinduism man is greater than the godheads he worships.[5] Rabindranath Tagore coined the happy expression "the humanity of God" long before Karl Barth used it.[6] Keshub Chunder Sen once called his brand of Hinduism "Human Catholic Religion."[7]

We in the West are suffering a malaise about the nature and worth of the human. Some changes must be made. We cannot adopt with any authenticity an Indian mode of life and thought, and we cannot realistically hope to attain a synthesis of all philosophical and religious traditions; but, after we have

127

given up the parochial assumption that all goodness and all truth originate in the West, we may be able to discover recuperative insights within Eastern traditions. Hinduism is a promising field in which to look for significant concepts of humanness.

WHAT IS NEO-VEDĀNTISM?

The title of this paper may be misleading as there is no school of thought named Neo-Vedāntism. The term designates certain trends within Hinduism during the last two centuries stimulated by the impact of the West and catalyzed by the Indian efforts to achieve independence from the British. This period of Indian history has been given various labels: "The Indian Renaissance," "The Age of Indian Awakening," and "The Indian New Age." However, these titles may suggest a more radical change than has in fact taken place. A revolutionary in India does not advertise himself as such, but he stresses his continuity with tradition. The first of the Neo-Vedāntists, Ram Mohun Roy, said his opposition was not to Brahmanism but to a perversion of Brahmanism. Nehru said Gandhi was the greatest of the revolutionists, but Gandhi said of himself, "I do not claim to have originated any new principle or doctrine. I have simply tried in my own way to apply the eternal truths to our daily life and problems. . . . I have nothing new to teach the world."[8] In India the innovator clothes his newness in the shrouds of the past. Winston Churchill, shocked by Gandhi's attire, once referred to "that half-naked fakir," but Gandhi adopted the simple loin cloth of the Indian peasant because he knew the villagers of India would never receive his message had he appeared in the black coat, grey tie, and striped trousers he had worn in London's Inns of Court.

The renaissance continues in India. The impacts of Western science, technology, education, social patterns, and democracy are sometimes striking. For example, the British have recently elected a woman as leader of the conservative party, but India has had a woman prime minister since 1966. India has joined the atomic bomb club—a dubious honor. Her resources in coal, iron, manganese, bauxite, and copper offer prospects for future industrial developments. India's humanitarianism was demonstrated during the Bangladesh war when she gave sanctuary to eight million refugees. Paul Ehrlich's suggestion in *The Population Bomb* that the principle of triage be applied to India, i.e., that she be denied assistance on the grounds that she will die regardless of treatment, may be premature. Moreover, as we in the West begin the necessary simplification of our lives, we may find profit in looking to the Indians for guidance inasmuch as some ecologists estimate each American pollutes as much air, land, and water as do eighty Indians.

The term Neo-Vedāntism is derived from *Veda* and *Vedānta*. The primary scriptures of Hinduism are called the *Vedas*. The word *Veda* is from the root *vid*, to know. In its narrowest sense Veda denotes the poetic petitions composed by priests to accompany the sacrifices to deified natural powers for health, wealth, progeny, victory in battle, immortality, etc. In a wider sense the term refers also to the priestly handbooks on rituals, to the allegorical interpretations of the rituals, and to the philosophical speculations known as the *Upaniṣads*. The *Upaniṣads* were composed a few centuries before and after the life of the Buddha. They do not present a single view but rather a wide variety of metaphysical and moral positions. Somewhere between 200 B.C. and A.D. 400 an intellectual named Bādarāyaṇa summarized what he regarded as the teachings of the *Upaniṣads* in 555 extremely pithy and practically unintelligible aphorisms which we know as the *Vedānta Sūtras*. The term Vedānta is composed of *Veda* (wisdom) and *anta* (end), so Vedānta means the summation of the ancient wisdom. The *Vedānta Sūtras* are perhaps best understood as mnemonic devices for students of the *Upaniṣads*. They were the original College Outline Series. These *Sūtras* required commentaries to explain their meaning. Although a wide variety of interpretations of the aphorisms of Bādarāyaṇa were created, only three were recognized as contenders for the honor of being the correct and final interpretation of the *Vedānta Sūtras*, and hence of the *Upaniṣads*. These were the Advaita (non-dualistic) school of Śaṅkara (ninth century), the Viśiṣṭādvaita (modified non-dualistic) school of Rāmānuja (eleventh century), and the Dvaita (dualistic) school of Madhva (thirteenth century). The fact that all three schools embody the term dualism in their name is a compliment to the first philosophers of India, the Sāṁkhya, who for unknown reasons developed the ontological dualism latent in the *Upaniṣads*, which must be described as a minority point of view. In the combat of philosophers since the ninth century Śaṅkara's non-dualism has been victorious. By the beginning of the nineteenth century the thinking associated with Śaṅkara and his followers was so dominant that the term Vedānta meant Advaita Vedānta, and if one wished to refer to Rāmānuja's or Madhva's systems one had to specify Viśiṣṭādvaita Vedānta or Dvaita Vedānta. This is still the situation among orthodox Hindus. In order to avoid confusion I shall in this essay refer to Śaṅkara's philosophy as Vedānta and to his supporters as Vedāntins and to the nineteenth and twentieth century reform and reformers as Neo-Vedāntism and Neo-Vandāntists.

I should add as a footnote to what has been said thus far that orthodox Hindus would modify this brief analysis by reminding us that in their opinion the *Vedas* have no human authors, that they are older than the world, that they are the original truth given orally to the ancient seers (*ṛsis*), and that therefore the *Vedas* are to be regarded as *śruti* (heard), a term which means a primordial revelation. Hence, for the orthodox the *Vedas* are unqualifiedly authentic and authoritarian. To this the objective scholar must reply that the *Vedas* must be interpreted by human minds.

The focus of Hinduism prior to the nineteenth century was individual salvation through transcendence. *Aham Brahman asmi* (I am Brahman) or *tat tvam asi* (That you are) was the existential confession which was sought. Advaita, Viśiṣṭā-dvaita, and Dvaita disagreed in some of the details of what this confession meant, but all agreed that each human being must be liberated from suffering and from the claims of the physical world. India's confrontation with the West challenged this view of man and the world. The British proved to be something new in the history of the sub-continent. For the first time the Hindus faced an invader whom they could not assimilate. Indeed, the Hindus found themselves in the role of the assimilated. Hindus began to look to the West. Pragmatic concerns challenged traditional transcendentalism. There came into existence, in the words of Thomas Babington Macauley, "a class of persons, Indian in blood and color, but English in taste, in opinions, in morals, and in intellect."[9] But the class we are considering did not go this far. They remained Hindus, although they were much affected by the West. The architects of this updated Hinduism I call the Neo-Vedāntists because they attempted to be faithful to the Vedānta tradition while assimilating Western ideas, values, and practices. The renaissance proceeded unevenly, for, as Gandhi says in his autobiography, "It is the reformer who is anxious for the reform, and not society."[10] In the villages life today goes on much as it has for centuries, but in the cities some Indians try to outdo Westerners in being Western, and others try to live in both worlds. Many business men in Bombay, New Delhi, Calcutta, and Madras wear Western coat and trousers at their offices and traditional Indian clothes in their homes.

The morning star of the Indian renaissance was Ram Mohun Roy (1772-1833). He broke from his Brahmin family, mastered Latin, Greek, and Hebrew in order to understand the Christian scriptures, founded the Brāhmo Samāj in Calcutta as a Hindu religion similar to Unitarianism, and among other interesting experiences converted a Protestant missionary to Hinduism. Roy was known as "the universal man" and as "the harbinger of the idea of Universal Humanism." Rabindranath Tagore said of him, "Through the great mind of Ram Mohun Roy the true spirit of India asserted itself and accepted the West, not by the rejection of the soul of India, but by the comprehension of the soul of the West."[11] Keshub Chunder Sen (1838-1884) made the Brāhmo Samāj still more Western, and finally split the movement, partly because of his undisguised zeal for Christianity. He sometimes referred to the Brāhmos as the "Hindu Apostles" of Christ.[12] Swami Vivekananda (1863-1902) translated the mystical experiences of his master, Ramakrishna, into programs of social welfare. He was the first Hindu to bring Hinduism to the United States. Rabindranath Tagore (1861-1941), the winner of the Nobel Prize for Literature in 1913, taught in his school at Śāntiniketan what might be called an international aesthetic Hinduism. His aim was to create individuals, not students. He wanted an education which would not kill the human by imposing the abstract. In 1921 he founded India's first coeducational university. Aurobindo Ghose (1872-1950) was arrested in 1907 for complicity in the use of bombs against the British, but he escaped imprisonment, due perhaps to his timely conversion to spiritual matters. In 1909 the British learned more about his activities and considered deporting him to Burma. But Aurobindo retreated to Pondicherry in South India, where he founded his ashrama, contending that the building of better human beings should precede efforts to drive the British from India. Mohandas Karamchand Gandhi (1869-1948) was the fashioner of Indian Independence. Gandhi is still an enigma, but few would deny that he belongs with the outstanding figures in human history. Sarvepalli Radhakrishnan (1888-1975) was the best-known Indian philosopher of the twentieth century. As professor at Oxford University, vice-chancellor of several Indian universities, vice-president and later president of India, and author of dozens of books, he interpreted Vedānta to the West and modified it in the direction of a universal religion. These are the leaders among the Neo-Vedāntists.

Others who could be listed include M. G. Ranade (1841-1901), B. G. Tilak (1856-1920), and G. K. Gokhale (1866-1915). These were linked in varying degrees to an offshoot of the Brāhmo Samāj in Bombay known as the Prārthanā Samāj. This society set for itself four objectives: to oppose the caste system, to introduce widow remarriage, to encourage female education, and to abolish child marriage. Tilak, who is called "the Father of Indian Unrest," is remembered for his vigorous opposition to Gandhi's policy of non-violence. Mention should also be made of Dayananda Saraswati (1824-1883), the founder of the Ārya Samāj, the communist leader M. N. Roy (1887-1954), Jawaharlal Nehru (1889-1964) whom Erik Erikson calls "intellectually the most brilliant statesman of our time,"[13] and the land reformer Vinoba Bhave (1895-   ). Each in his distinctive manner contributed to a pattern of humanness which has significance for us in the West as we raise the old and yet ever new question "What does it mean to be human?"

NEO-VEDĀNTISM AND HUMAN KNOWLEDGE

Śaṅkara held a scholastic view of knowledge. All reliable knowledge comes ultimately from the *Upaniṣads*: "The comprehension of Brahman is effected by the ascertainment, consequent on discussion, of the sense of the Vedānta-texts, not either by inference or by the other means of right knowledge. . . . Scripture itself, moreover, allows argumentation; for [certain] passages . . . declare that human understanding assists scripture."[14] Again he wrote, "Brahman . . . cannot become an object of perception; . . . inference also and the other means of proof do not apply to it; but, like religious duty, it is to be known solely on the ground of holy tradition."[15] Reasoning was allowed only as a subordinate auxiliary to the knowledge gained by faithfully receiving the *śruti* literature.

The Neo-Vedāntists broke from this view of knowledge chiefly because of their exposure to Western forms of rational-empirical methodologies. One of the reasons why Dayananda Saraswati cannot be fully classified with the Neo-Vedāntists is because he was, if possible, more authoritarian than Śaṅkara.

According to Dayananda the *Vedas* contain all the knowledge human beings have and all they will ever have. The keys to the solution of all social and political problems are hidden in the *Vedas*. By the most incredible exegesis Dayananda discovered adumbrations of modern science and technology in these ancient writings. For example, the line "I invoke Mitra and Varuna for the success of my poem"[16] was interpreted as a statement that water is the combination of hydrogen and oxygen! Aurobindo, while holding the Vedic tradition, in high esteem, admitted that "the age of intuitive knowledge, represented by the early Vedāntic thinking of the *Upaniṣads*, had to give place to the age of rational knowledge; inspired Scripture made room for metaphysical philosophy, even as afterwards metaphysical philosophy had to give place to experimental Science."[17] Keshub Chunder Sen with characteristic enthusiasm went to the other extreme: "Science will be your religion. . . . In the New Faith everything is scientific."[18]

The first step in the Indian renaissance was the breaking of the stranglehold of the *Upaniṣads* and Śaṅkara's interpretation of them. Devendranath Tagore, the father of Rabindranath, early broke from Vedānta on theistic grounds: "Those *Upaniṣads* which treated Brahma were alone accepted by us as the true Vedānta. We had no faith in the Vedānta philosophy because Śaṅkaracharya seeks to prove therein that Brahma and all created beings are one and the same. What we want is to worship God."[19] In 1823 the British government appropriated funds for the promotion of education in India, but there was no agreement as to what should be the nature of this education. Some contended it should be exclusively Oriental studies; others held it should be chiefly Western science and technology. When the "Orientalists" appeared to be winning, Ram Mohun Roy wrote a long letter to William Pitt in which he said that he had been filled with sanguine hopes that the instruction would be in "Mathematics, Natural Philosophy, Chemistry, Anatomy and other useful Sciences, which the Nations of Europe have carried to a degree of perfection that has raised them above the inhabitants of other parts of the world."[20] Ram Mohun continued, "We now find that the Government are establishing a Sanskrit school under Hindu pundits to impact such knowledge as is already current in India. This seminary . . . can only be expected to load the minds of youth with grammatical niceties and metaphysical distinctions of little or no practical use to the possessors or to society. . . . The Sanskrit language so difficult that almost a lifetime is necessary for its perfect acquisition is well known to have been for ages a lamentable check on the diffusion of knowledge. . . . Nor will youths be fitted to be better members of society by the Vedāntic doctrines, which teach them to believe that all visible things have no real existence."[21]

Gandhi was the Neo-Vedāntist least qualified to have reliable opinions in epistemological matters, yet he wrote the most in this area. He cared little for scriptural authority (*śruti-pramāna*). He denied the exclusivity of the *Vedas*: "I believe the Bible, the Quran, and the Zend Avesta to be as divinely inspired as the *Vedas*."[22] His eclecticism was not a cultural smorgasbord, for he remained a Hindu: "I do not want my house to be walled in on all sides and my windows to be stuffed. I want the cultures of all lands to be blown about my house as freely as possible. But I refuse to be blown off my feet."[23] He also said, "All that is printed in the name of scriptures need not be taken as the word of God or the inspired word. But every one cannot decide what is good and authentic and what is bad and interpolated. There should be some authoritative body that would revise all that passes under the name of scriptures, expurgate all the texts that have no moral value or are contrary to the fundamentals of religion and morality and present such an edition for the guidance of Hindus."[24] This was an amazing suggestion from Gandhi, because he constantly pontificated what was good and authentic in the Hindu scriptures, and he would have been the first to reject the decisions of an expurgative committee. The great body of moral writings, the *Dharma Śāstras*, he said would be "death traps" were Hindus to regulate their conduct according to all the details given in them.[25] Gandhi's own hermeneutical principles were three. Some of the moral prescriptions he rejected as "later interpolations." For example, when asked if he approved of the doctrine of *varna* as prescribed in the *Laws of Manu*, he replied, "The principle is there. But the applications do not appeal to me fully. There are parts of the book which are open to grave objections. I hope they are later interpolations."[26] Sometimes he said that moral precepts found in the scriptures which are "contrary to known and accepted morality . . . must be rejected."[27] In addition to using "later interpolations" and "accepted morality" as principles of interpretation, he also rejected all that is "in conflict with sober reason or the dictates

of the heart."[28] When Gandhi was asked about the prescriptions
for early marriage in the *Śāstras*, he replied, "We must reject
them in the light of positive experience and scientific
knowledge."[29] "Even a Vedic text," he said, must be rejected "if
it is repugnant to reason and contrary to experience."[30] But we
must ask, "Whose reason? Whose experience?" The answer can
only be "Gandhi's." In one of his characteristic emphatic
moods Gandhi wrote, "Even if all the Hindus of India were to be
raged against me in declaring that untouchability, as we know
it today, has the sanction of *śāstras* and *smṛtis*, I will declare
that these *śāstras* and *smṛtis* are false."[31] The final authority
for Gandhi was the inner voice: "The still small voice within
you must always be the final arbiter when there is a conflict of
duty."[32] Usually his moral intuitions were sound, but if later
experience proved otherwise, he was the first to acknowledge
that he had made a mistake. Gandhi was a pragmatist. He said
and did what he thought the occasion demanded. Sometimes he
acted with surprising disregard for the feelings of those nearest
and dearest to him. His inconsistencies were many. In 1933 he
wrote in self-defense, "To a diligent student of my writings and
to those who are interested in them, I do want to point out that I
hardly care for appearing to be the same at all times. I have
given up many cherished thoughts and have learnt a number of
new things. I have grown in age but I do not think my spiritual
evolution has ended or is likely to end even after my
death. . . . Hence whoever finds inconsistency as between my
two writings should consider the latter of the two to be
authentic if he has any faith in my wisdom."[33] Early in his
career Gandhi said "God is Truth," but later he changed to
"Truth is God," claiming that there is a "fine distinction
between the two statements."[34] Inasmuch as he also said
"Truth . . . is but another name for God,"[35] I am at a loss to
determine the distinction. Sometimes his pragmatism—he
referred to himself as a "practical idealist"[36]—led to relativism:
truth became that which "presents itself to me at a given
moment."[37] When Ralph Coniston of *Collier's Weekly* asked
Gandhi what he would advocate if he were to attend the
meetings in San Francisco which established the United
Nations, he replied, "If I knew I would tell you but I am made
differently. When I face a situation, the solution comes to me. I
am a man of action. I react to situations intuitively. Logic
comes afterwards, it does not precede the event."[38] He was a
strange man, a man of many contradictions. Gunnar Myrdal

says he "established a pattern of radicalism in talk but conservatism in action that is still very much a part of the Indian scene."[39] He was a man of action and a man of prayer, a humble seeker after truth who could be arrogantly dictatorial, a tender-hearted man who was often callous of other peoples' feeling, a Hindu whom the Christian missionary E. Stanley Jones called "one of the most Christlike men in history."[40]

THE HUMAN IN A REAL WORLD

According to Śaṅkara we human beings are afflicted with a pernicious delusion which he called *adhyāsa*. This is the tendency to superimpose the characteristics of one thing on another thing. *Adhyāsa* appears principally in the application of the immediate and obvious characteristics of the world of appearance to the world of reality. We assume the surface of things is the reality of things. These forms of ignorance alienate us from the Brahman, the integral ground of being, consciousness, and value which is best expressed in the Sanskrit term *Satchitānanda*. Vedānta has a rigorous conception of reality. The real must be absolutely, independently, and necessarily real. The real is the self-existent. This is the sort of reality for which Anselm argued in his ontological proofs for the existence of God, i.e., a being whose essence is existence. *Asat* (non-being), the opposite of *sat* (being), is that which cannot be because its existence would involve a contradiction. The usual example is the son of a barren woman. The world of sense experience according to the Vedāntins lies between *sat* and *asat*. This is the world of *māyā*. This is the world of space and time, of name and form, of dependence upon the Brahman. The word *māyā* is usually translated "illusion," although this tends to lead to the notion that the world of our sense experience is unreal. Thus the Vedāntins seem to be saying that there are two kinds of unreality: absolute unreality (*asat*) and relative unreality (*māyā*). A better translation of *māyā* is "appearance." This is more consonant with the historical fact that Śaṅkara's doctrine of *māyā* was supported by appeal to the term *iva* (as it were). The traditional example of a *māyā* is a "snake" experienced by a weary traveler at twilight when he first sees in the path what proves upon examination to be a rope. The object of the traveler's alarm is not an illusion; it is a rope-identified-as-a-snake. The illusion is not the object in the path. The illusion is the error of apprehension made by the traveler.

The Neo-Vedāntists either rejected or radically modified

Śankara's doctrine of *māyā*. They repudiated the Vedāntic devaluation of the spatial and temporal world.

Vivekananda said the doctrine of *māyā* calls our attention to the relative value of the world and of human life in that world. To call the world *māyā* means that the world is full of contradictions. He said, "The world is an indefineable mixture of reality and appearance, of certainty and illusion."[41] He also said, "The world has neither existence nor non-existence. You cannot call it existent because that alone truly exists which is beyond time and space, which is self-existent. Yet this world does satisfy to a certain degree our idea of existence. Therefore we can say that it has an apparent existence."[42]

Rabindranath Tagore was very original in his reinterpretation of *māyā*. He approached the doctrine aesthetically: "The world as an art is *māyā*. It is and is not. Its sole explanation is that it seems to be what it is. The ingredients are elusive; call them *māyā*, even disbelieve them, the Great Artist, the Māyāvin, is not hurt."[43] More philosophically, Tagore held that the doctrine of *māyā* is the means by which we falsely claim that the world is real in its own right, that it has nothing to do with the Infinite. *Satya* (true reality) reveals the interdependence of the Infinite and the world: "Without the world, God would be phantasm; without God, the world would be chaos," wrote Tagore in "Stray Birds." Tagore argued that although some philosophers deny that there is any finitude, i.e., that the finite is but *māyā* (illusion), for him the finite is real and "the word *māyā* is a mere name, it is no explanation. It is merely saying that with truth there is this appearance which is the opposite of truth; but how they come to exist at one and the same time is incomprehensible."[44] Here, as in many other places in his writings, Tagore affirms his belief in metaphysical polarity: finite and infinite, male and female, good and evil, positive and negative, God and man. He lacked the philosophical mind to develop this significant insight, or at least he as poet did not choose to develop it. Often one has the feeling that Tagore was more Taoist than Hindu, for something like a Yin-Yang polar duality seems to underlie much of his writing, e.g., he says in *Sādhanā* that "the world in its essence is a reconciliation of pairs of opposing forces. These forces, like the left and right hands of the creator, are acting in absolute harmony, yet acting from opposite directions."[45] And in *Naibedya*, Poem 88 he writes, "This will I admit: how the One became two I know not, nor ever shall know."

Radhakrishnan was the Neo-Vedāntist most sympathetic to

Vedānta. Therefore his criticism of Śaṅkara's doctrine of *māyā* was restrained. He explicitly denied that the doctrine means the empirical world and individual selves are illusions.[46] He claimed that the doctrine was established to account for the "fragility of the universe."[47] He thought the ancient sages of India were more impressed by the Heraclitean aspects of the world than by the Parmenidean: "The feature of the world which led the Hindu thinkers to raise the question of the Real was its passing away."[48] Radhakrishnan stressed the moral rather than the ontological implications of the doctrine of *māyā*. This is consistent with his philosophical idealism which is rooted in ethical ideals rather than in metaphysical or epistemological ideas. Thus he found a pragmatic purpose in the doctrine: to induce humans to transfer their attention from the finite and transitory to that which is the foundation of all values, all realities, and all truths.

Aurobindo totally rejected the doctrine of *māyā*. He referred to it as "the tyrannous falsehood of *māyā*."[49] His rejection was based on his view that if the world and the individual self are illusions then man's liberation and his part in the evolution from matter to the highest forms of spirit are also illusions. He would have nothing to do with a gnostic form of liberation: "*Sādhana* has to be done in the body, it cannot be done by the soul without the body."[50] He ridiculed the Śaṅkara doctrine of *māyā*: "Therefore we arrive at the escape of an illusory non-existent soul from an illusory non-existent bondage in an illusory non-existent world as the supreme good which that non-existent soul has to pursue!"[51] Aurobindo drove a wedge between himself and the Vedāntins when he stated that the world-negation which prevails in India is due to the teachings of the Buddha which were taken up and completed by Śaṅkara.[52] Nothing angers the Vedāntins more than to hear Śaṅkara called a Buddhist. Aurobindo's efforts to eliminate the feelings of unreality, frustration, and disappointment which he found in India earned for him the approbrium "The Liberator." We experience a real world, he said, although we may at times understand it incorrectly. We must stop thinking in terms of illusions and dreams. We must get on with our proper human responsibility of "conscious evolution"[53] to the realization of the God-man which is our potentiality. Aurobindo called human beings "to become complete in being, in consciousness of being, in force of being, in delight of being and to live in this integrated completeness."[54] He reminded his readers over and over again that however high we human individuals climb we must not forget that our roots are in matter. We must not abandon the lower to itself, but must transfigure it in the light of the higher. In a delightful essay entitled "Perfection of the Body" Aurobindo argued that the aim is always "a divine life in a divine body." The material remains active and effective at all levels in the evolutionary schema of Aurobindo. His use of the term *māyā* is a return to the original meaning of power, although for Aurobindo it is the power of the One to become many and of the many to become One, of spirit and matter to interact.

Jawaharlal Nehru was the least metaphysical of all the Neo-Vedāntists, yet he caught well the new emphasis of *māyā* as a form of ontological relativity. Thus, in *The Discovery of India* he wrote, "So we find in India, as elsewhere, these two streams of thought and action—the acceptance of life and the abstention from it—developing side by side, with the emphasis on the one or the other varying in different periods. Yet the basic background of that culture was not one of other-worldliness or world-worthlessness. Even when in philosophical language, it discussed the world as *māyā*, or what is popularly believed to be illusion, that very conception was not an absolute one but relative to what was thought of as ultimate reality . . . and it took the world as it is and tried to live its life and enjoy its manifold beauty."[55]

THE HUMAN IN A UNIFIED WORLD

The Vedāntins claim that the universe cannot be understood if we think in terms of dualisms or pluralisms. Therefore, they call their philosophical system "Advaita" (non-dualism). The universe, according to them, is such a unity that there can be no philosophy of anything unless it is a philosophy of everything. The part cannot be known separate from the whole. We do not know, and maybe will never know, how this cosmos came into being, say the Vedāntins, but we cannot deny that beings have appeared, consciousness has appeared, and values have appeared. Reality is like that. All that is, all that is conscious, and all that is valuable is rooted and grounded in the Totality which is Brahman or *Satchitānanda*. Tagore has expressed this as follows: "The meaning is that Reality, which is essentially one, has three phases. The first is *Sat*; it is the simple fact that things are, the fact which relates us to all things through the relationship of common existence. The second is *Chit*; it is the fact that we know, which relates us to all things through the relationship of knowledge. The third is *Ananda*; it is the fact that we enjoy, which unites us with all things through the relationship of love."[56]

The Neo-Vedāntists do not disagree with the monistic thrust of Vedānta, but they think of monism as internal relations, as a unity of coherent parts, rather than as absolute totality. They stress that the Absolute cannot be totality without the existence of individuals.

This conception of a unified world means that the distinction between the inanimate and the animate cannot be sharp. C. P. Raman's researches in the biological growth of crystals and J. C. Bose's studies of the affective life of plants are two examples of the impact of this conception on Indian scientists.

The human oneness with nature is expressed at Tagore's schools in classes held in a mango grove. The children are encouraged to go barefoot, and the ashramites often walk to the edge of Śāntiniketan before the evening meal to enjoy the sunset. The evening of full moon the students parade around the compound singing Tagore's songs. Tagore once said that the best place to read a book on botany is the crotch of a tree. He wrote in an essay on his school, "We come to this world to accept it, not merely to know it. We may become powerful by knowledge, but we attain fullness by sympathy. The highest education is that which does not merely give us information but makes our life in harmony with all existence."[57] Tagore dreamed of a unified world as a world of unified humanity. His university, he said, is an international university, a universal university, where all will find their place.

Vivekananda translated the concept of a unified world into its meaning for our attitude toward death. "The fear of death," he said, "can only be conquered when man realizes that so long as there is one life in this universe, he is living. When we can say 'I am in everything, in everybody, I am in all lives, I am the universe,' then alone comes the state of fearlessness."[58]

For all the Vedāntists the unity of the world means that no line can be drawn between ethics and politics, or between religion and politics, or between religion and economics. Gandhi once wrote, "I could not be leading a religious life unless I identified myself with the whole of mankind and that I could not do unless I took part in politics."[59] The following words of Gandhi appear over the entrance to the Gandhi Museum in Delhi: "I am told that religion and politics are different spheres of life. But I would say without a moment's hesitation and yet in all modesty that those who claim this do not know what religion is." Ram Mohun Roy goaded the British into making and enforcing laws eliminating child marriage, polygamy, and the burning of widows despite the British desire not to interfere with the moral and religious customs of the Hindus.

Gandhi also interpreted the unified world as the oneness of the world of life. He wrote, "Hinduism insists on the brotherhood not only of all mankind but of all that lives. It is a conception which makes one giddy, but we have to work up to it."[60] And elsewhere he wrote, "Men cannot be really good or really civilized unless they can embrace in their goodness or their civilization all that lives."[61] During his life he experimented with the application of unity in two related areas: nonviolence and diet. His principle of nonviolence or harmlessness (ahiṁsā) has often been misunderstood. Perhaps the most persistent misunderstanding is that it is the refuge of the weak and the cowardly. But Gandhi said, "I do believe that where there is a choice only between cowardice and violence, I should advise violence."[62] Ahiṁsā, according to Gandhi, is the principle of strong persons and groups who refuse to use weapons of violence. It is not a simple principle, and the application is very complex, for our lives are caught up in violence whatever be our intent. Gandhi explained, "Ahiṁsā is a comprehensive principle. We are helpless mortals caught in a conflagration of hiṁsā. The saying that life lives on life has a deep meaning in it. Man cannot for a moment live without consciously or unconsciously committing outward hiṁsā. The very fact of his living—eating, drinking and moving about— necessarily involves some hiṁsā, destruction of life, be it ever so minute. A votary of ahiṁsā therefore remains true to his faith if the spring of all his actions is compassion, if he shuns to the best of his ability the destruction of the tiniest creature, tries to save it, and thus incessantly strives to be free from the deadly coil of hiṁsā. He will be constantly growing in self-restraint and compassion, but he can never become entirely free from outward hiṁsā."[63] Throughout his life Gandhi tried to reconcile the inner principle of ahiṁsā with the unavoidability of outward hiṁsā. He also struggled to find a diet which would square with his belief in the oneness of life. Life should not feed on life—but how can this be avoided? The youthful Gandhi promised his mother he would not eat meat, and the vow

bothered him throughout his life as he tried at various times to
be a vegetarian, or an ovo-lactarian, or a granivore, or a
fruitarian. At one time he limited himself to peanut butter and
lemons. At the time of his death he was eating only fruits and
nuts. His book, *The Story of My Experiments with Truth* could
almost be titled *The Story of My Experiments with Food*. The
pursuit of Truth by nonviolent means was the heart of
Hinduism for Gandhi, and this was grounded in his conviction
of the unity of life.

THE HUMAN IN AN ORDERED WORLD

The Neo-Vedāntists claimed that Hinduism has no conflict
with the sciences. The basis for this claim is the Hindu doctrine
of cosmic order. From early times the emphasis has been on the
regularity and uniformity of nature rather than on the power of
divinities to violate natural order. The term *ṛta* was used by the
ancient seers to refer to the alternation of day and night, the
rotation of the seasons, and the movement of the heavenly
bodies. By reason of the Hindu belief in the unity of the cosmos,
ṛta also denoted the moral order which is manifested in the
consequences of righteous and unrighteous behavior. *Ṛta* was
not regarded as fate or the will of God but as the law of cause
and effect operating in both natural and moral phenomena.

The *rta* of morality is usually called *dharma*, a term
frequently badly translated as "duty," "obligation," "responsibility," or "religion." But *dharma* is fundamentally an
ontological concept referring to the essence of things. *Dharma*
is what a thing does which shows forth its nature. The *dharma*
of fire is to burn, of water to make things wet. So when the
dharma concept is applied to the rational animal it is not
surprising that there is a dual meaning: there are both
descriptive *dharmas* and prescriptive *dharmas*, both what one
does and what one ought to do. The human as a physical body
falls to the center of the earth, and as a living being ingests,
digests, egests, grows, and reproduces. The human as a
rational being ought to think clearly, logically, creatively, and
critically; the human as a social being ought to function well as
a member of a family, a community, and a nation; and the
human as a spiritual being ought to appreciate and contribute
to the totality of goodness, beauty, and truth. *Dharma* also
refers to individual natures, and thus has profound implications in the so-called caste system. For example, the
dharma of a Brahmin is to study and to teach the *Vedas*. The

Brahmin who fails to continue Vedic studies throughout his life is a Brahmin in name only, a crypto-Brahmin, not a true Brahmin.

The Neo-Vedāntists capitalized on the doctrine of *dharma*. Gandhi did not use the term often, but he constantly reminded Indians of their obligations to eliminate untouchability, to get the British out of India, to meliorate the relations between Muslims and Hindus, and to support themselves through the work of their own hands. Tagore wrote, "*Dharma* is the ultimate purpose that is working in our self. When any wrong is done we say that *dharma* is violated, meaning that the lie has been given to our true nature."[64] Self-development to becoming the Universal Person was a *dharma* he laid upon everyone. Aurobindo argued that the human is potentially the Superhuman and that each individual should enter into conscious evolution toward that goal. Radhakrishnan similarly stressed the *dharma* of growth: "Man's final growth rests with himself. His future is not solely determined, like that of other animals, by his biological past. It is controlled by his own plans for the future.[65] *Dharma* denoted the fixity of human nature for the Vedāntins, but it denoted the capacity for growth, change, and development for the Neo-Vedāntists.

The ṛta principle was also interpreted for centuries in India in the doctrine of *karma* (the law of the deed) and in the consequent doctrine of *saṁsāra*. *Saṁsāra* is usually translated "reincarnation," or "transmigration," or "birth-death-rebirth," but its broader meaning is the condition of being subject to the ravages of time. *Karma* and *saṁsāra* have been accepted as basic beliefs in Hinduism since at least the sixth century B.C. Louis Renou has said that "the gods are really superfluous in Indian religion. . . . [The] essentials could have been covered by the theory of *karma* and its consequences."[66] *Karma* is causality applied to morality. It is the conviction that one reaps only what one sows, and that what one sows one will surely reap. The doctrine of reincarnation appears to have developed to account for those instances in which either the sowing or the reaping cannot be identified in the span of years between birth and death. Some sowing may have been done in a previous existence, and some reaping may be done in a future existence. Another motivation for the development of the doctrine of reincarnation may have been the almost universal conviction of the conscious animal to believe that death is not the absolute ending of life. It is illuminating to note that the first word to designate this conviction in Hinduism is *punarmṛtyu* (the death of death). The ancient *ṛṣis* appear to have concluded that both the states of life and death are temporally limited. The end of life is called death, so the end of death is called life. They spoke of redeath, rather than rebirth. Indeed, the term rebirth is most inappropriate as it arouses the image of a ghost moving from one body to another.

The Neo-Vedāntists made few changes in the doctrines of *karma* and *saṁsāra*. They accepted them as part of the cultural assumptions within which they worked. For the most part they played down the imagery, affirming their confidence that the death of the human individual does not terminate opportunities for growth and avoiding any discussion of the mechanics of reincarnation. Aurobindo's statement that "not only the elements of our physical body, but those of our subtler vital being, our life-energy, our desire-energy, our powers, strivings, passions enter both during our life and after our death into the life-existence of others"[67] is a fine expression of their view of *saṁsāra*.

Radhakrishnan examined *karma* in terms of the freedom-determinism controversy so familiar to Western philosophers. Although he is no more successful in solving the puzzle than are his Western counterparts, one of his similes is interesting. He observes that the human life is like a game of cards: "The cards in the game are given to us; we do not select them. They are traced to past *karma*, but we are free to make any call we like and to lead any suit. Only, we are limited by the rules of the game. We are more free when we start the game than later on, when the game has developed and our choice becomes restricted. But till the very end there is always a choice."[68] This interpretation, however, is not new in Hinduism, for in the *Bhagavad Gītā* Lord Kṛṣṇa tells Arjuna that, while he cannot avoid acting, the manner in which he acts is under his control.

The doctrines of *karma* and *saṁsāra* imply the inequality of human beings. No two individuals have the same *karma*. No two have attained the same degree of self-realization. This has interesting consequences for education, social structures, and government in India. The Hindu citizen of the modern nation of India, while recognizing that political equality is right and just, does not confuse this with human inequalities. Hence, in India a man can usually discuss the inequalities of individuals without being accused of being a fascist, a nazi, a snob, or a male chauvinistic pig.

IMPORTANCE OF THE HUMAN INDIVIDUAL

One of the most important changes in the renaissance brought about by the Neo-Vedāntists is the new view of the reality and worth of the human individual. Śaṅkara contended that the real Self is non-dual consciousness with no attributes. His argument for the reality of only one Self—the *Ātman*-Brahman—sprang from his notion of the absoluteness of the subjectivity of the *Ātman*. The Sāṁkhya philosophers held that the object must be the object of a subject, and that the subject must be the subject of an object. But Śaṅkara held that the *Ātman* is absolute subjectivity—a unity which transcends the subject-object relationship. According to him the individual self (*jīva*) is an instance of relative subjectivity. The notion of an individual self arises from identifying the body with the Self. This individuality is like the yellowness of a piece of white marble seen through amber. Individuality is a *māyā* attribute of the real. The individual self has no ontological status. It is only a phenomenal aspect of Reality. Hence, the individual self is of little worth.

The Neo-Vedāntists completely reject this conception of the human individual. They repudiate the Vedāntic claim that the Ātman shines forth of itself. They insist that the individual human exists for the noble purpose of manifesting *Satchitānanda*. For them the classic Upanisadic formula "*Ātman* is Brahman" does not establish the equivalence of *Ātman* and Brahman; rather it affirms that each *ātman* (*n.b.* lower case *a*) is a monad which mirrors the Absolute. They hold that the self is subject only in the presence of an object, and hence that the *ātman* cannot be identified with Brahman. "Individuality is universality in the plan of creation," declared Vivekananda.[69]

Another approach to this important point may be helpful. By starting from the principle of unity rather than from the principle of universality, the Neo-Vedāntists state that if Brahman is absolute reality, then the world and the individual are not external to Brahman but a sort of internal dimension. Relativity is an aspect of the Absolute. Brahman would not be Totality were pluralization and individualization omitted. No unfulfilled possibility can be excluded from Brahman.

Brahman unpluralized would not be Brahman. The human self is Brahman pluralized into individuals. Humans share existence as living beings with all plants and lower animals. This is their nature as *jīvas*. But in addition humans have a dimension of selfhood not shared with plants and lower animals. This is their self-awareness. This is their nature as *ātmans*. The human is the being who encounters himself as a being. In this internal conceptualizing the human constitutes himself as agent. The human ability to know the self is the basis of the moral consciousness and the possibility of liberation. Only an *ātman* can become aware of the unique relation of itself to the whole. Only an *ātman* can launch on the path of perfecting to *Satchitānanda*.

The Neo-Vedāntists refer again and again to the worth and dignity of the individual. Keshub Chunder Sen in a lecture entitled "Great Men" said, "For it must be admitted that every man is, in some measure, an incarnation of the divine spirit."[70] Aurobindo refers approvingly to a section in the *Mahābhārata* in which the fire god Agni says, "I am present in the bodies of men as vital fire."[71] Aurobindo argues that the best form of social service is personal growth: "True, his life and growth are for the sake of the world, but he can help the world by his life and growth only in proportion as he can be more and more freely and widely his own real self."[72] Gandhi stressed the individual in his political thinking: "I look upon an increase of the power of the state with the greatest fear, because, although apparently doing good by minimizing exploitation, it does the greatest harm to mankind by destroying individuality which lies at the root of all progress."[73] But the socialist Nehru was more cautious, warning that India must strike a balance between "the old Hindu conception of the group being the basic unit of organization, and the excessive individualism of the West, emphasizing the individual above the group."[74]

Tagore was the Neo-Vedāntist who had the most to say about the worth of the individual. In his book *My Reminiscences* he says that the single theme which dominated his life and thought is that "the great is to be found in the small, the infinite within the bounds of form."[75] He speaks of "the joy of attaining the Infinite within the finite."[76] His biographer, Edward Thompson, recounts a conversation in his last years when Tagore said, "It is strange that even when so young I had that idea, which was to grow with my life all along, of realizing the Infinite in the finite, and not, as some of our Indian metaphysicians do, eliminating the finite."[77] Our individuality, he writes in another essay, is the only thing we can call our own, and if this is lost, it is a loss to the whole world, and hence he says, "The desire we have to keep our uniqueness intact is really the desire of the universe acting in us."[78] Although Tagore believed, "The whole weight of the universe cannot crush out this individuality of mine,"[79] yet he constantly warned against the pressures of conformity.

Tagore believed that women are the custodians of this precious emphasis on the individual: "A man's interest in his fellow-beings becomes real when he finds in them some special gift of power or usefulness, but a woman feels interest in her fellow-beings because they are living creatures, because they are human, not because of some particular purpose they can serve, or some power which they possess and for which she has a special admiration."[80] But Tagore felt that women did not fully appreciate what they had to contribute, and he warned, "At the present stage of civilization, when the mutilation of individuals is not only practised, but glorified, women are feeling ashamed of their own womanliness. For God, with his message of love has sent them as guardians of individuals. . . . But because men in their pride of power have taken to deriding things that are living and relationships that are human, a large number of women are screaming themselves hoarse to prove they are not women, that they are true where they represent power and organization."[81] Tagore, therefore, offered the following advice to each woman: "She must restore the lost balance by putting the full weight of the woman into the creation of the human world. The monster car of organization is creaking and growling along life's highway, spreading misery and mutilation, for it must have speed before everything else in the world. Therefore woman must come into the bruised and mained world of the individual; she must claim each one of them as her own, the useless and the insignificant. . . . The world with its insulted individuals has sent its appeal to her. These individuals must find their true value, raise their heads once again in the sun, and renew their faith in God's love through her love."[82]

THE HUMAN AS MĀRGAYĀTA

There is a curious bifurcation in Hindu culture: life affirmation and life denial, sensuality and asceticism, excess and deficiency, beauty and ugliness, tradition and creativity.

Aurobindo's shift from political activism to reclusion is an example. Vinoba Bhave shifted in the opposite direction. When he was a student of religious literature at Banaras, he was torn between the desire to seek asylum in the Himalayas and the desire to devote himself to freeing India from the British. Gandhi's message of attaining self-realization through service to humanity resolved the conflict. He joined Gandhi's movement in 1916, and in 1940 became the first disciple to engage in civil disobedience. Since Gandhi's death he has branched out in his own movements: *bhūdān* (gift of land), *sampattidān* (gift of wealth), and *jīvandān* (gift of self). Gokhale was "a rationalist who believed in astrology, a statesman who founded a monastic society, a poet who taught mathematics . . . [a] responsible administrator and [an] external agitator."[83] Erik Erikson says he senses in Gandhi's autobiography "the presence of a kind of untruth in the very protestation of truth; of something unclean when all the words spelled out an unreal purity; and, above all, of displaced violence where nonviolence was the professed issue."[84] The tension within human nature is stressed by the Neo-Vedāntists in their analysis of man as a becoming. The human being is ever on the way—rising-falling, growing-shrinking, advancing-receding—yet ever on the move. The human being is human only when becoming something more. Nandalal Bose has captured this view beautifully in his linocut of the walking Gandhi, the eternal pilgrim, staff in hand, forward gait, never looking back. The human being is a *mārgayāta*, a wayfarer. His state is a state of becoming, not of being. The gods are—man becomes. "To become is our life's significance," said Aurobindo.[85] Radhakrishnan said that the human being is "a transitional being, an unfinished experiment."[86] Again he said, "The human being is a *saṁsarin*, a perpetual wanderer, a tramp on the road. His life is incessant metamorphosis."[87] His one ideal is "to make himself profoundly human, perfectly human. . . . Self-discovery, self-knowledge, self-fulfilment is man's destiny."[88] Nehru, however, cautioned, "Perfection is beyond us for it means the end, and we are always journeying, trying to approach something that is receding."[89] Vivekananda put it more theologically: "Man is to become divine, realising the divine more and more from day to day in an endless progress."[90] And Tagore expressed the same idea more fully: "Man is not complete; he is yet to be. In what he *is* he is small, and if we could conceive of him stopping there for an eternity we should have an idea of the most awful hell that man can imagine. In his *to be* he is infinite, there is his heaven, his deliverance. His *is* is occupied every moment with what it can get and have done with; his *to be* is hungering for something that is more than can be got, which he never can lose because he never had possessed."[91] The title of Gandhi's autobiography—*The Story of My Experiments with Truth*—correctly indicates the Neo-Vedāntists' view of the human life. Gandhi refused to write a thesis on *ahimsā*, maintaining that he was always experimenting with it and he did not wish to give the appearance of completion. On one occasion he said, "Let us be sure of our ideal. We shall ever fail to realize it, but shall never cease to strive for it."[92] He wrote, "There must, therefore, be ceaseless striving after perfection. . . . Infinite striving after perfection is one's right. It is its own reward."[93] The continual development of which the Neo-Vedāntists speak is linked with the belief in *karma* and *samsāra*, as Gandhi indicated when he wrote that "if for mastering the physical sciences you have to devote a whole lifetime, how many lifetimes may be needed for mastering the greatest spiritual force—non-violence—that mankind has known?"[94] If the Neo-Vedāntic doctrine of human development sounds like the old Christian heresy of Pelagianism, we need to keep in mind that the perfecting of which they speak is measured over many incarnations.

This is one of the places where the Vedāntins and the Neo-Vedāntists differ significantly in their conceptions of humanness. Whereas the Vedāntins cling to a closed system of life based on the authoritativeness of the *Upaniṣads*, the Neo-Vedāntists hang loose to life, refusing to capsule the human in any tidy and finished system. Their thinking is always provisional and open-ended.

The Neo-Vedāntists appear to be the first to be aware fully of the polarity of the goals in Hinduism. On the one hand the Hindu tradition stresses *ātmansiddha*, the perfection of the true self. This embodies self-restraint, self-denial, and even self-abuse. This is the tradition stereotyped in the common Western image of a *sadhu* (holy man) lying on a bed of spikes. There is no denying that such has existed and does exist in India. The word *āśrama* which is used for institutions associated with spiritual retreats, education, and meditation comes from a root meaning to suffer, which would seem to imply that there is no progress unless there is suffering. I saw this illustrated in an elementary school south of New Delhi which had this motto over the entrance: "No pains, no gains." But on the other hand the Hindu tradition put emphasis on *nāgaraka*, which can only be interpreted as libertarianism. The prescriptions for securing hedonic satisfactions in some of the literature of Hinduism surpass even Lord Chesterfield's letters to his son. The techniques of kissing, biting, scratching, and copulating are exhaustively catalogued in the *Kāma Sūtra*. Whereas Hindus prior to the Neo-Vedāntists argued for one or the other of the goals—for *ātmansiddha* or for *nāgaraka*—the Neo-Vedāntists attempted to reconcile the two. The human being—that strange and wonderful being poised between the animal and the divine—is to seek development of divine potentialities and yet not to neglect his animality. Gandhi said he found the entire message of Hinduism in the first verse of the *Īśa Upaniṣad*. His words were that "if all the *Upaniṣads* and all the other scriptures happened all of a sudden to be reduced to ashes, and if only the first verse of the *Īśa Upaniṣad* were left intact in the memory of Hindus, Hinduism would live forever."[95] He translated the first verse into four *mantras*: "All that we see in this great universe is pervaded by God. Renounce it. Enjoy it. Do not covet anybody's wealth or possession." Renounce-enjoy! One is to enjoy life without becoming possessed by the enjoyment. Tagore hinted at the possibility of a harmony of human salvation and human happiness: "For us the highest purpose of this world is not merely living in it, knowing it and making use of it, but realising our own selves in it through expansion of sympathy; not alienating ourselves from it and dominating it, but comprehending and uniting it with ourselves in perfect union."[96]

Neo-Vedāntists interpret Hinduism as a common quest, not as a common faith. The quest is outlined for the Hindu in what is known as *sādhana*. This term defies translation, although it can be described as a total discipline prescribed for human beings for the approximation of ideal goals. Hindu *sādhana* has been spelled out through the centuries in four ways. One is the *chaturvarga* (tetrad of goods), the blessings which no human ought to miss. They are *kāma* (hedonic delights), *artha* (material possessions), *dharma* (fulfillment of obligations to the communities of which humans are members: the joint family, the civic community, the animal kingdom, the world of scholarship, and the ancestral spirits), and *mokṣa* (liberation from the claims of the embodied life). Another aspect of *sādhana* is the dividing of the human span of years into four divisions, each with its special joys and responsibilities: student, householder, retiring person, and *sannyāsi*. Just as the student period is the preparation for the period of supporting and perpetuating the community, so the retirement period is a preparation for the final period in which full attention is given to the attainment of *mokṣa*. Another way to think of the four periods is that the first is the period of academic learning, the second two are periods of learning through practical experience, and the final is the second period of scholarship. Aurobindo has described the last period as follows: "When you have paid your debt to society, filled well and admirably your place in its life, helped its maintenance and continuity and taken from it your legitimate and desired satisfactions, there still remains the greatest thing of all. There is still your own self, the inner you, the soul which is a spiritual portion of the Infinite, one in its essence with the Eternal. This self, this soul in you you have to find, you are here for that."[97] A third aspect of *sādhana* is the *mārgas*. These are four optional life styles or paths (*mārgas*) to enlightenment. If the Hindu selects wisely, he will follow a path consistent with his aptitudes and talents. The four *mārgas* are *jñāna mārga* (the path of knowledge), *karma mārga* (the path of works), *bhakti mārga* (the path of devotion to a god), and *yoga mārga* (the path of physical and psychological disciplines). The Hindu is assured under the principle of *iṣṭa mārga* (chosen path) that any one conscientiously followed will lead to the goal. The fourth aspect of *sādhana* is the class system of Hindu society. Scholars are far from agreement as how the distinction between Brahmins, Kṣatriyas, Vaiśyas, and Śūdras came to be, and also as how a large group of Hindus came to be excluded from the four classes.

There are two features of the class system which must be distinguished in order to understand the reactions of the Neo-Vedāntists. One of these is *varṇa*. This is the proper term for the four classes. *Varṇa* denotes families that have descended from the class of priests and scholars (Brahmins), from the class of rulers and warriors (Kṣatriyas), from the class of merchants and traders (Vaiśyas), and from the class of farmers and manual laborers (Śūdras). The other feature of the class system is *jāti*. This denotes further divisions based for the most part on vocations and guilds. There are between two and three thousand *jātis*. Although one often hears Hindus say that both *varṇa* and *jāti* are determined by birth, a more correct statement would be that *karma* determines *varna* and birth determines *jāti*. *Varṇa* refers to the attainments of the *mārgayāta*. A Brahmin is not a Brahmin because he is born into a Brahmin family; rather he is born into a Brahmin family because he is a Brahmin by reason of the *karma* of his previous existences. But whether he is a Brahmin priest, or a Brahmin scholar, or a Brahmin cook is determined in the orthodox tradition by his *jāti*, i.e., if his family is a family of cooks, then he will probably become a cook. While the *varṇa* and *jāti* classifications have served useful purposes in Hindu society, they have also corrupted it by the hundreds of prescriptions and prohibitions which have been added to *varṇa* and *jāti*. For example, according to the *Dharma Śāstras* a Brahmin upon meeting a fellow Brahmin is supposed to inquire about his spiritual well-being, upon meeting a Kṣatriya to ask if he is free from pain, upon meeting a Vaiśya to ask if has lost anything, and upon meeting a Śūdra to ask if he is free from disease. There are also prescriptions for the height of the salute of greeting, the length of the *nim* stick used for cleaning the teeth, the amount of interest to be paid on borrowed money, etc. Many of the *jāti* rules deal with customs governing eating, drinking, social intercourse, and marriage. The problem for the liberal-minded and innovative Neo-Vedāntist was—and is—how to preserve the values of the system and to eliminate the defects. Vivekananda finally lost all patience with the whole structure and called it a "crystallized social institution, which after doing its service is now filling the atmosphere of India with stink."[98] Likewise Aurobindo said that the whole social structure "has become a name, a shell, a sham and must either be dissolved in the crucible of an individualist period of society or else fatally affect with weakness and falsehood the system of life that clings to it."[99] Tagore enjoyed the benefits which accrued to him by reason of his Brahmin *varṇa*, but as a humanist he rebelled against the inequities which had been added. He wrote, "Among the doctrines of the new age that come to us is the one that makes all men equal before the law. Whether a Brahmin kills a low-caste Śūdra or a Śūdra kills a Brahmin, it is murder all the same, and calls for the same punishment. No ancient injunction in this regard can sway the scales of justice."[100]

Nehru was surprisingly vague about the Hindu class system. He may have felt that enough alteration was taking place in Indian society without disrupting this part of traditional social behavior. He wrote, "It has ceased to be a question of whether we like caste or dislike it. Changes are taking place in spite of our likes or dislikes. But it is certainly in our power to mold those changes and direct them, so that we can take full advantage of the character and genius of the Indian people as a whole, which have been so evident in the cohesiveness and stability of the social organization they built up."[101]

Gandhi supported *varṇa*. It is "inherent in human nature."[102] It is "the best form of insurance of happiness."[103] It is "the law of man's being and therefore as necessary for Christianity and Islam, as it has been for Hinduism."[104] Anyone who does not live according to the *varṇa* into which he is born is "a degraded being."[105] Gandhi also wanted to preserve the *jāti* system insofar as it was a system of guilds. He held that a person ought to follow the vocation of his family, although the person might engage in other activities as avocations. Gandhi repudiated untouchability as a malignancy of *varṇa*, and he contended that when untouchability was eliminated *varṇa* would be purified. B. R. Ambedkar, the chairman of the committee which wrote the Indian Constitution and himself an Untouchable, disagreed. He held that Gandhi's solution was simply to raise all Untouchables to the level of Śūdras. Gandhi's stand on *varṇa* and *jāti* was often puzzling, but at one point he was clear and helpful: he contended that human beings have equal civil rights and unequal social responsibilities.

THE HUMAN IN SOCIETY

Although it would be an exaggeration to say that the nineteenth century marked the dawn of social conscience in India, the Neo-Vedāntists did put a new emphasis on the social. Hindu culture had always stressed the importance of natural units such as the family and the clan, but the Neo-Vedāntists emphasized volunteer organizations. The term *samāj* (society) became popular, as in Brāhmo Samāj, Brāhmo Samāj of India, Ādi Brāhmo Samāj, Sādharaṇ Brāhmo Samāj, Nava Bidhava Samāj, Ārya Samāj, and Prārthanā Samāj. Several of the Neo-Vedāntists established ideal communities. Keshub Chunder Sen established Bharat Ashram at Belgharia seven miles north of Calcutta in 1872. It was a communistic society divided into four groups on the basis of whether the individual wished to stress wisdom, love, service, or *yoga* in his spiritual discipline. The community lasted only five years. Tagore established Sāntiniketan in 1901. It began as a school for boys, but it has expanded into a cluster of schools and colleges, village service centers, and homes for scholars, artists, poets, and philosophers. Gandhi out of his experiences with the Phoenix Settlement and the Tolstoy Farm in Africa established Sabarmati near Ahmedabad in 1915 and Sevagram near Wardha in 1932. Both welcomed Untouchables to full membership. Vinoba Bhave founded a model city named Paunar. Perhaps the most ambitious of all the communities of the Neo-Vedāntists is Aurobindo's ashram at Pondicherry. On February 28, 1968 a model city, a "city of human unity and universal culture," was launched at Pondicherry with the blessing of the United Nations. At the inaugural ceremony the universality of Auroville was symbolized by young people of one hundred and twenty countries who poured a bit of earth from each of their countries into a lotus-shaped urn.

The Neo-Vedāntists learned from the British the importance of citizenship. Lionel Smith said in the House of Commons in 1831 that the introduction of Western learning into India would make the Indians discover the importance of governing themselves, and Thomas Macauley in his speech on Indian education (1835) predicted that the proudest day in English history would be the day when the Indians demanded European institutions. The Neo-Vedāntists were among the first Indians to argue for Indian nationhood. The Indian National Congress was established in Bombay on December 28, 1885. Ranade, Tilak, and Gokhale were leaders in the early days. Gandhi first attended a meeting of the Congress in 1901, and it was he more than anyone who changed the Congress from a middle-class debating society into a mass organization for promotion of Indian independence and nationalism.

Meanwhile, the Neo-Vedāntists were stressing the impor- tance of social service. The *Dharma Śāstras* are filled with admonitions of responsibilities to members of the joint family, of the community, and to a lesser degree of the kingdom. The so- called Golden Rule is neatly expressed in the *Mahābhārata*: "Do naught to others which, if done to you, would cause you pain. This is the sum of *dharma*."[106] The Vedāntic philosophers reminded Hindus that each should love his neighbor, for in the final analysis one's neighbor is one's self. Vivekananda, in a manner characteristic of the Neo-Vedāntists, contended that "the religion of the future enlightened humanity" is developing in India, a religion in which one "looks upon and behaves to all mankind as one's own soul." He lamented that it "was never developed among the Hindus universally."[107] The Ramakrishna-Vivekananda movement which he organized in 1897 has developed into a worldwide collection of schools, colleges, social centers, adult education projects, libraries, hostels, and hospitals. Gandhi's work among the outcaste, whom he called Harijans (Children of God) continues in India today in the form of free elementary education, cottage industries, university scholarships, and reserved seats in the national parliament.

I have indicated that there is some question whether Dayananda Saraswati can be classified with the Neo-Vedāntists because of his traditional scholastic attitude toward the *Vedas*. But his progressiveness with respect to social issues puts him with the Neo-Vedāntists. He organized the Ārya Samāj with perfect equality between the sexes. He attempted to eliminate untouchability, even engaging in missionary endeavors to induce Untouchables to join the Ārya Samāj. He angered orthodox Hindus by investing Untouchables with the sacred thread. His followers continue to do excellent work in their schools, hospitals, orphanages, and community centers. They are sometimes called "The Aggressive Hindus." They are a perfect example of the kind of people Radhakrishnan had in mind when he wrote, "The world is in dreadful need of these heroic spirits who have the courage of their vision of human oneness to assume the new leadership."[108]

The Neo-Vedāntists evaluate the state in terms of the quality of the lives of its citizens which it fosters rather than in terms of the gross national product. Nehru wrote in what proved to be among his last printed words as part of the forward to a volume of letters edited by his friend Shriman Narayan, "Socialism has become rather a vague word. . . . But . . . we must not forget that the essential objective to be aimed at is the quality of the individual and the concept of *dharma* underlying it."[109] Tagore warned, "I do not put my faith in any institutions, but in the individuals all over the world who think clearly, feel nobly, and act rightly."[110] While one might suppose that Gandhi, because of his training as a lawyer, would have been favorable to the state, the truth is that he did not hide his conviction that the ultimate test of the state is its value for the lives of individuals. He wrote in *Young India*, August 13, 1925, "I remain loyal to an institution so long as that institution conduces to my growth." This was the basis for his civil disobedience. He swore that everything he did was for self-realization. It is not widely known that Gandhi was fundamentally an anarchist. He wished to limit the functions of the state to the narrowest possible limits. He, like Tagore, Aurobindo, and Bhave, urged people to turn away from the national state to the village. *Pañchāyata rāj* (village self-government) is still discussed as an ideal in India, although at the same time most people look to New Delhi for solutions for their problems. Gandhi hoped for the withering away of the state and the evolving of what he called "benign autocracy." He wrote in *Young India*, July 2, 1931, "There is a state of enlightened anarchy. In such a state everyone is his own ruler. He rules himself in such a manner that he is never a hindrance to his neighbour. In the Ideal State there is no political power because there is no State. But the ideal is never fully realized in life. Hence the classical statement that that government is best which governs the least."

The nation is a puzzling feature of Indian history. During the early part of this century, when India was trying to become independent from the British, some of the leaders of this movement had doubts about the wisdom of becoming a nation at the very time in which they were struggling for nationhood. For example, Muhammad Ali Jinnah said at the First Roundtable Conference in London, "God made man and the Devil made the nation."[111] The concept of nation did not come easily to a people with such varied backgrounds. "We have no word for 'Nation' in our language," wrote Tagore in a letter to C. F. Andrews. "When we borrow this word from other people, it never fits us."[112] The confusion is illustrated by a resolution passed by the Indian Communist Party in September 1942 that every section of the Indian people with "common homeland, common tradition, common language, common culture, common psychological make-up, and common economic life" should be a nation.[113] This resolution could be interpreted to mean that there should be hundreds of nations in South Asia or that there should be no nation whatsoever!

Vivekananda used his oratorical skills to inspire young men to become heroes in the liberation movement. He probably would have been surprised and disappointed had he known that his words would be used after his death to stimulate militant nationalistic movements in Bengal, for he had reservations about the national ideal: "When a man has reached the highest, when he sees neither man nor woman, neither sex, nor creed, nor colour, nor birth, nor any of these differentiations, but goes beyond and finds that divinity which is the real man behind every human being—then alone he has reached the universal brotherhood."[114]

Tagore believed that Gandhi's non-cooperative movement would foster a feeling of isolation among Indians, while he believed that a feeling of the unity of mankind should be encouraged. Tagore speculated that the human being moves through three stages: the individual, the national, and the complete or universal. One of the worst things that could happen would be for a people to level off at the national stage. He wrote, "The complete man must never be sacrificed to the patriotic man."[115] In a letter to Leonard Elmhirst the poet condemned what he called "short-sighted nationalism,"[116] and he wrote to C. F. Andrews that the goal should not be *swarāj* (self-rule) but the emancipation of man from the meshes of national egoism.[117] What Tagore had in mind was a universal humanism, a union of the peoples of the world based on understanding and love. He wanted united peoples, not united nations.

On August 15, 1947, when the Indians were rejoicing in their freedom from British rule, Aurobindo issued an Independence Day Declaration in which he asked the Indians to look beyond mere nationalism. He wrote that "international forms and institutions must appear, perhaps such developments as dual or multilateral citizenship, willed interchange or voluntary fusion of cultures. Nationalism will have fulfilled itself and lost its militancy and would no longer find these things incompatable with self-preservation and the integrality of its outlook. A new spirit of oneness will take hold of the human race."[118]

India's non-alignment stance since becoming a nation may be an expression of some of these ideas. In the light of the reservations expressed by the Neo-Vedāntists we might begin asking if the game of nations is becoming too expensive and too dangerous to play. Some nation must speak out for the human against the inhumanness of nations. Some nation must restore confidence in the importance of individual human beings. In the words of Radhakrishnan, "Self-perfection is the aim of religion, but until that aim takes hold of society as a whole, the world is not safe for civilization and humanity."[119]

The Neo-Vedāntists have made still another contribution to

the concept of human sociality. This is the belief that there is no individual salvation, for the salvation of the individual presupposes the salvation of others. The ideal individual and the perfect society arise together. No one, however enlightened, can be liberated until all are liberated. This Neo-Vedāntic view is very similar to the Buddhist conception of the *bodhisattva*. The view was expressed by B. K. Goswami, a co-worker of Keshub Chunder Sen, as follows: "It is selfishness to tread the solitary way to righteousness. We must enter the Kingdom of Heaven taking all with us."[120] Radhakrishnan called this *sarvamukti* (universal salvation).

NEO-VEDĀNTIC FAITH IN THE HUMAN

When twelve million Hindus, Muslims, and Sikhs were moving in contrary directions in the Punjab after Independence, killing each other in their efforts to flee either into India or into Pakistan, a message came to Gandhi to come at once. He replied, "Never lose faith in humanity." How often this has been repeated by the Neo-Vedāntists! Gandhi knew well the evil in the human, but he wrote, "Man's nature is not necessarily evil. Brute nature has been known to yield to the influence of love. You must never despair of human nature."[121] He spoke often of the infinite possibilities of the individual to develop nonviolence. He never ceased to encourage men and women to experiment in the ways of peace.

The optimism of the Neo-Vedāntists sometimes stands out against the pessimism of Westerners. In 1939 when we in the West were talking about lights going out in Europe, Radhakrishnan wrote, "In spite of appearances to the contrary, we discover in the present unrest the gradual dawning of a great light, a convening life-endeavour, a growing realization that there is a secret spirit in which we are all one, and of which humanity is the highest vehicle on earth, and an increasing desire to live out this knowledge and establish a kingdom of spirit on earth."[122] Radhakrishnan conjectured that we in the West may be passing through a new renaissance. Our first renaissance resulted from a rebirth of the classical culture of ancient Greece and Rome. Our second may be effected by the discovery of India. The result will be the creation of a new homeland of the spirit in which we shall understand one another. Radhakrishnan as he wrote must have had in mind the last line of the *Ṛg Veda*: "Common be your intention; common be the wishes of your hearts; common be your thoughts, so that there may be thorough union among you."[123]

I wish I could share this optimism, but I think of Lila Roy, an American woman married to an outstanding Bengali novelist, who wrote in 1955 that in India "a new element has emerged, an element which reflects the essential unity of the human mind and gives rise to a broader humanism than the world has ever before sought to practice."[124] Mrs. Roy recently visited the United States and her home state of Texas after an absence of more than a quarter of a century, but she found life here so intolerable that she cut her stay short and returned to India.

On May 7, 1941, the day of his eightieth birthday and exactly three months before his death, Rabindranath Tagore delivered a heartbroken address entitled "Crisis in Civilization." The outbreak of war in Europe had saddened him almost beyond endurance. He said, "I had at one time believed that the springs of civilization would issue out of the heart of Europe. But today when I am about to quit the world that faith has gone bankrupt altogether. . . . And yet I shall not commit the grievous sin of losing faith in Man. . . . A day will come when unvanquished Man will retrace his path of conquest, despite all barriers to win back his lost heritage."[125] He closed his address with a poem written especially for the occasion.

> The Great One comes
> sending shivers across the dust of the Earth.
> In the heavens sounds the trumpet,
> in the world of man drums of victory are heard,
> the Hour has arrived of the Great Birth.
> The gates of Night's fortress
> crumble into the dust—
> on the crest of the awakening dawn
> assurance of a new life
> proclaims "Fear not."
> The great sky resounds with hallelujahs of victory
> to the Coming of Man.[126]

This poem must have reminded Tagore and his audience of an earlier poem of his in which he expressed his abiding confidence in the human being:

> There on the crest of the hill
>> stands the Man of Faith amid the snow-white
>>> silence.
>> He scans the sky for some signal of light,
>> and when the clouds thicken and the nightbirds
>>> scream as they fly,
>> he cries, "Brothers, despair not, for Man is great."[127]

NOTES

1. *Ṛg Veda* 10. 90. 13-14, Ralph T. H. Griffith translation. In this essay I have taken the liberty of modifying the transliteration of some Sanskrit terms in the interest of consistency. Consonants are pronounced as in English, and vowels as in German. Note that *ś* is pronounced as in *"show,"* ṣ as in *"church,"* and ṛ as in *"reed."*

2. *Santi Parva*, Section 300, *The Mahabharata*, P. C. Roy edition (Calcutta: Orient Publishing Co., n.d.), vol. 9, p. 404.

3. A. Chakravarti, "Humanism and Indian Thought," *Miller Lectures* (Madras: University of Madras, n.d.), p. 27.

4. *Young India*, Nov. 13, 1924.

5. Aurobindo, "The Indian Conception of Life," in *Silver Jubilee Commemoration Volume* of the Indian Philosophical Congress, 1950, p. 174.

6. Rabindranath Tagore, *Man* (Allahabad: Kitabistan, 1946), p. 162.

7. Quoted from Keshub Chunder Sen, "Tracts for the Times" by Prosanto Kumar Sen, *Biography of a New Faith* (Calcutta: Thacker, Spink and Co., 1950), vol. I, p. 250.

8. M. K. Gandhi, *Hindu Dharma*, ed. Bharatan Kumarappa (Ahmedabad: Navajivan Publishing House, 1950), p. 3.

9. Thomas Babington Macauley, "Minute on Education" (1830), *Sources of Indian Tradition*, ed. William Theodore de Bary (New York and London: Columbia University Press, 1958), vol. II, p. 49.

10. *Gandhi's Autobiography: The Story of My Experiments With Truth*, trans. Mahadev Desai (Washington, D.C.: Public Affairs Press, 1948), p. 268.

11. Quoted by Edward Thompson, *Ethical Ideals in India Today* (London: Watts and Co., 1942), p. 23.

12. Erik Erikson, *Gandhi's Truth: On the Origins of Militant Nonviolence* (New York: W. W. Norton and Co., 1969), p. 265.

13. de Bary, *Sources of Indian Tradition*, vol. II, p. 74.

14. *The Vedānta Sūtras With the Commentary by Śaṅkarākārya*, 1. 1. 2, trans. George Thibaut, ed. Sarvepalli Radhakrishnan and Charles E. Moore, *A Source Book of Indian Philosophy* (Princeton: Princeton University Press, 1957), p. 511.

15. *Ibid.*, 2. 1. 6, p. 522.

16. *Ṛg Veda* 1. 2. 7.

17. Aurobindo, *The Life Divine* (New York: The Greystone Press, 1949), p. 65.

18. From a pamphlet entitled *Epistle to Indian Brethren* written in 1880.

19. *The Autobiography of Maharashi Devendranath Tagore*, trans. Satyendra Nath Tagore and Indira Tagore (Calcutta: S. K. Lahiri and Co., 1909), p. 24.

20. Sophia Dobson Collet, *The Life and Letters of Raja Rammohun Roy*, 3rd ed. (Calcutta: Sadharan Brahmo Samaj, 1962), p. 458.

21. *Ibid.*, pp. 458-59.

22. *Hindu Dharma*, p. 7.

23. *Young India*, June 1, 1921.

24. *Hindu Dharma*, p. 384.

25. *Ibid.*, p. 21.

26. *Ibid.*, p. 329.

27. *Ibid.*, p. 354.

28.   Quoted by K. P. Karunakaran, *New Perspectives on Gandhi* (Simla: Indian Institute of Advanced Study, 1970), p. 26.

29.   *Ibid.*, p. 358.

30.   *Ibid.*, p. 37.

31.   Karunakaran, *New Perspectives on Gandhi*, p. 36.

32.   *Young India*, August 4, 1920.

33.   Quoted by Maganbhai P. Desai, "Gandhi's Way of Life," in *Quest for Gandhi*, ed. G. Ramachandran and T. K. Mahadevan (New Delhi: Gandhi Peace Foundation, 1970), pp. 71-72.

34.   *Young India*, Dec. 31. 1931.

35.   *Gandhi: Essential Writings*, ed. V. V. Ramana Murti (New Delhi: Gandhi Peace Foundation, 1970), p. 414.

36.   V. S. Naravane, *Modern Indian Thought* (Bombay: Asia Publishing House, 1964), p. 184.

37.   *Ibid.*, p. 183.

38.   Pyarelel, *Mahatma Gandhi: The Last Phase* (Ahmedabad: Narajivan Publishing House, 1956), vol. I, p. 120.

39.   Gunnar Myrdal, *Asian Drama* (New York: Random House, 1968), vol. II, p. 756. One student of Indian life writes with respect to the Congress Party, "Indian politics which during the struggle for freedom had been largely a genuine expression of sacrifice became instead an avenue for profit. There is nothing particularly surprising in this. Even genuine revolutions inevitably give way to bourgeois values but India was and still is to some extent unique. The range of radical social and economic policies continually endorsed by the dominant party and ignored in principle is unequalled elsewhere." Michael Edwardes, *Nehru: A Political Biography* (New York and Washington: Praeger Publishers, 1971), pp. 249-50.

40.   Naravane, *Modern Indian Thought*, p. 173.

41.   In an address he gave to the Philosophical Society of Harvard University on March 25, 1896, *Ibid.*, p. 93.

42.   *Ibid.*

43.   Rabindranath Tagore, "The Religion of an Artist," in *Contemporary Indian Philosophy*, ed. Sarvepalli Radhakrishnan and J. H. Muirhead (London: George Allen and Unwin, 1936), p. 38.

44.   Rabindranath Tagore, *Sādhanā* (New York: The Macmillan Co., 1914), p. 95.

45.   *Ibid.*, p. 96.

46.   Sarvepalli Radhakrishnan, *Eastern Religions and Western Thought* (London: Oxford University Press, 1939), p. 28.

47.   *Ibid.*

48.   Sarvepalli Radhakrishnan, *The Heart of Hindusthan* (Madras: G. A. Nateson and Co., 1945), p. 1.

49.   *The Life Divine*, p. 670.

50.   Aurobindo, *Bases of Yoga* (Calcutta: Arya Publishing House, 1949), p. 134.

51.   *The Life Divine*, p. 39.

52.   *Ibid*, p. 374.

53.   *Ibid.*, p. 936.

54.   *Ibid.*, p. 909.

55.   Jawaharlal Nehru, *The Discovery of India*, ed. Robert I. Crane (Garden City: Doubleday and Co., 1960), p. 46.

56. Rabindranath Tagore, *Creative Unity* (London: Macmillan and Co., 1959), p. 48. Cf. Lewis Thomas, *The Lives of a Cell: Notes of a Biology Watcher* (New York: Viking, 1974). Thomas likens the planet Earth to a single cell.

57. Rabindranath Tagore, *Personality* (London: Macmillan and Co., 1959), p. 116.

58. *The Complete Works of Swami Vivekananda* (Calcutta: Advaita Ashram, 1958), vol. II, pp. 80-81.

59. M. K. Gandhi, *Non-violence in Peace and War* (Ahmedabad: Navajivan Publishing House, 1951), vol. I, p. 170.

60. *Hindu Dharma*, p. 35.

61. *Collected Works of Mahatma Gandhi* (New Delhi: Publications Division, Ministry of Information, 1958), vol. I, p. 89.

62. A statement made by Gandhi on August 25, 1920. Quoted by Louis Fischer, ed, *The Essential Gandhi* (New York: Random House, 1962), p. 156. Although I have avoided making any evaluation of Gandhi's principle of nonviolence in the body of my paper, I feel that I must express my conviction that he was in fact a man of violence. I regard his constant overriding of the wishes of other people as acts of violence. He was confident that he knew what was best for others—for his wife, his sons, the members of his ashrams, the Congress Party, and India—and the happiest moments of his life seemed to have been those in which he was meddling in the lives of other people. Michael Edwardes contends that Gandhi's fasts were "blackmail," (*Nehru: A Political Biography*, pp. 55, 152, 232) and I agree, despite his claim that he did not fast. against anyone but for the purification of his own soul. There is a pattern within Hinduism of exerting pressure upon another by threatening to do violence to one's self, and it is within this context that Gandhi's threats to fast and Nehru's threats to resign from public office must be examined.

63. *Gandhi's Autobiography*, pp. 427-28.

64. *Sādhanā*, p. 74.

65. Sarvepalli Radhakrishnan, "The Ancient Asian View of Man," in *Man's Right to Knowledge* (New York and London: Columbia University Press, 1954), p. 11.

66. Louis Renou, *Religions of Ancient India* (London: Athlone Press, 1953), p. 68.

67. *The Life Divine*, p. 186.

68. Sarvepalli Radhakrishnan, *An Idealistic View of Life* (London: George Allen and Unwin, 1952), p. 279.

69. *The Complete Works of Swami Vivekananda*, vol. VI, p. 121.

70. Quoted by P. K. Sen, *Keshub Chander Sen* (Calcutta: The Art Press, 1938), p. 60.

71. *Adi Parva*, Section 7.

72. Aurobindo, *The Human Cycle* (Pondicherry: Sri Aurobindo Ashram, 1949), p. 80.

73. *Selections from Gandhi*, ed. Nirmal Kumar Bose (Ahmedabad: Narajivan Publishing House, 1948), p. 27.

74. Nehru, *The Discovery of India*, p. 148.

75. Rabindranath Tagore, *My Reminiscences* (London: Macmillan and Co., 1917), p. 237.

76. *Ibid.*, p. 238.

77. Edward Thompson, *Rabindranath Tagore: Poet and Dramatist*, 2nd ed. (London: Oxford University Press, 1948), p. 51.

78. *Sādhanā*, p. 70.

79.   *Ibid.*, p. 69.

80.   *Personality*, p. 175.

81.   *Ibid.*, pp. 178-79.

82.   *Ibid.*, p. 182.

83.   Stanley A. Wolpert, *Tilak and Gokhale: Revolution and Reform in the Making of Modern India* (Berkeley: University of California Press, 1962), p. 25.

84.   Erikson, *Gandhi's Truth*, p. 231.

85.   *The Life Divine*, p. 901.

86.   Sarvepalli Radhakrishnan, *The Brahma Sūtra: The Philosophy of Spiritual Life* (London: George Allen and Unwin, 1960), p. 153.

87.   Sarvepalli Radhakrishnan, "Fragments of a Confession" in *The Philosophy of Sarvepalli Radhakrishnan*, ed. Paul Schilpp (New York: Tudor Publishing Co., 1952), p. 47.

88.   *Eastern Religions and Western Thought*, p. 35.

89.   Nehru, *The Discovery of India*, p. 684.

90.   *The Complete Works of Swami Vivekananda*, vol. I, p. 332.

91.   *Sādhanā*, p. 153.

92.   In a speech delivered in September, 1917. M. K. Gandhi, *Speeches and Writings*, 4th ed. (Madras: S. Ganesan, n.d.), p. 363.

93.   *Gandhi's Autobiography*, p. 114.

94.   *Harijan*, March, 1936.

95.   *Hindu Dharma*, p. 37.

96.   *Creative Unity*, p. 49. Krishna Kripalani says that Tagore and Gandhi "knew that the Indian *sādhanā* or the way of realising the truth of life was two-fold, through *tapasya* or penance and sacrifice, and through *ānanda*, a joyous acceptance of the fullness of life as partaking of the divine." He quotes a letter from Tagore in which the poet says that "Gandhi is the prophet of *tapasya* and I am the poet of *ānanda*." See "What I Owe to Tagore and Gandhi," *Indian Horizons* 23, no. 1 (1974) :45.

97.   Aurobindo, *The Foundations of Indian Culture* (New York: The Sri Aurobindo Library, 1953), p. 130.

98.   Quoted by K. M. Panikkar, *Hindu Society at Cross Roads* (Bombay: Asia Publishing House, 1955), p. 77.

99.   *The Human Cycle*, p. 11.

100.   Rabindranath Tagore, *Towards Universal Man* (Bombay: Asia Publishing House, 1961), p. 344.

101.   Nehru, *The Discovery of India*, p. 149.

102.   *Hindu Dharma*, p. 333.

103.   *Ibid.*, p. 328.

104.   *Ibid.*, p. 324.

105.   *Ibid.*, p. 326.

106.   Quoted by Monier Monier-Williams, *Indian Wisdom* (London: William H. Allen and Co., 1875), p. 446.

107.   In a letter to Mohammad Sarfaraz Husain of Naini Tal dated June 10, 1898. *The Complete Works of Swami Vivekananda*, vol. VI, p. 415.

108.   Sarvepalli Radhakrishnan, *The Religion We Need* (London: Ernest Benn Ltd., 1928), p. 32.

109.   Shriman Narayan, *Letters from Gandhi, Nehru and Vinoba* (New York: Asia Publishing House, 1968), p. 134.

110.   *Creative Unity*, p. 148.

111.   Quoted by Sankar Ghose, *Socialism, Democracy and Nationalism in India* (Bombay: Allied Publishers, 1973), p. 101.

112.   *Sources of Indian Tradition,* vol. II, pp. 239-40.

113.   Sankar Ghose, *Socialism, Democracy and Nationalism In India,* p. 180.

114.   *The Complete Works of Swami Vivekananda,* vol., I, pp. 390-91.

115.   Quoted by Krishna Kripalani, *Rabindranath Tagore* (London: Oxford University Press, 1962), p. 290.

116.   S. L. Malhotra, *Social and Political Orientations of Neo-Vedantism* (Delhi: S. Chand and Co., 1970), p. 100.

117.   *Ibid.,* p. 101.

118.   See Sisirkumar Mitra, *The Liberator: Sri Aurobindo, India and the World* (New Delhi: Jaico Publishing House, 1954), p. 190.

119.   *The Religion We Need,* p. 30.

120.   See P. K. Sen, *Keshub Chander Sen,* p. 112.

121.   *Harijan,* November, 1938.

122.   *Eastern Religions and Western Thought,* p. 3.

123.   *Rg Veda* 10. 191. 4, H. H. Wilson translation.

124.   Lila Roy, *Equities* (Bangalore: Indian Institute of Culture, 1955), p. 76.

125.   *Faith of a Poet: Selections from Rabindranath Tagore,* ed. Sisirkumar Ghose (Bombay: Bharatiya Vidya Bhavan, 1964), p. 55.

126.   *Ibid.,* p. 56.

127.   *The Viśvabharati Quarterly,* vol. 26, nos. 3 and 4:48.

INDEX